Screening the World

Stuart Hanson

# Screening the World

Global Development of the Multiplex Cinema

Stuart Hanson
De Montfort University
Leicester, Leicestershire, UK

ISBN 978-3-030-18994-5          ISBN 978-3-030-18995-2   (eBook)
https://doi.org/10.1007/978-3-030-18995-2

Cover Image: The Point multiplex in Milton Keynes. Colin Palmer Photography / Alamy Stock Photo
Cover Design: eStudio Calamar

This Palgrave Macmillan imprint is published by the registered company Springer Nature Switzerland AG
The registered company address is: Gewerbestrasse 11, 6330 Cham, Switzerland

*For Vidyasara*

# ACKNOWLEDGEMENTS

I am indebted to many people who have helped me in the course of writing this book—in particular Karsten Grummitt from Dodona Research, who generously furnished me with much data. Nancy Gallagher helped me with a great deal of useful material about AMC and made me most welcome when I visited Kansas City. Kam Dosanjh from Vue Cinemas has been a great source of knowledge and expertise. Judith Thissen and Karina Aveyard looked at relevant sections of the book and their comments were a great help. Mike Walsh from Flinders University very helpfully supplied me with a range of materials and sources about cinemas in Australia. Last but not least, James Chapman's extremely supportive comments on the finished manuscript were a tremendous vote of confidence.

I would also like to thank my colleagues in the Leicester Media School at De Montfort University and the Cinema and Television History Research Institute in particular.

Finally, I would like to thank my wife Vidyasara for her love and support.

# CONTENTS

# CHAPTER 1

# Introduction

This book examines the development of the multiplex from its beginnings in the USA in the early 1960s to its expansion overseas from the mid-1980s, across Europe and into parts of Asia-Pacific, including Australia. More than a consideration of overseas expansion on the part of US companies, the book considers the hegemony of the multiplex as a cultural and business form. It argues for the significance of the multiplex cinema as a phenomenon that has transcended national boundaries, and which has become the predominant venue for film viewing globally. In many countries, indigenous exhibition companies have adopted the multiplex template and adapted it to their own domestic markets. Finally, this book will be concerned with understanding how the development of the multiplex cinema has been shaped by broad social, economic, cultural, and political contexts, whilst avoiding an over-deterministic focus on simply technology. Implicit in this analysis is a recognition of the domination of US media multinationals and Hollywood cinema, and the development of the multiplex cinema as symbolic of the extension and maintenance of US cultural and economic power. Writing in 1992, Larry Gleason from Paramount Pictures argued that the:

> decade of the '90s will see the multiplex cinema strengthen its position as the single most innovative concept in the last 25 years in the world of exhibition. It has caused a rebirth and resurgence of cinema-going throughout the world. Starting in the U.S. in the late '60s, this phenomenon has in the last five years spread throughout the world, from Europe to Australia.[1]

© The Author(s) 2019
S. Hanson, *Screening the World*,
https://doi.org/10.1007/978-3-030-18995-2_1

1

Before we go any further though, we need to ask: what *is* a multiplex? In 1973, Sumner Redstone, owner of National Amusements, was mulling over what to call one of his sixplex cinemas that scheduled more films in a week than it had the number of screens, largely as a way of making it harder for distributors to monitor how many films were showing. "The word plex was in the lexicon and I worked with that" he said, and then "it came to me. 'Multiplex!' I jotted down the word down and said it out loud. That's what we had, a multiplex."[2] Redstone then sought legal advice as to whether the term could be trademarked and he discovered it could be, so, as he recalled, "National Amusements now owns the trademark to the word "multiplex." Although it has become common worldwide, when you see the word "multiplex" attached to a theatre in the United States, it belongs to National Amusements."[3]

For Redstone, as we have seen, it was, pragmatically, a cinema with six screens or more. In India, such was the complexion of the existing single-screen cinema infrastructure, a multiplex is any modern cinema with more than one screen.[4] Nonetheless, the term began to be used in the US trade press in the early 1970s, to describe both the subdividing of an existing cinema into multiscreen venues and the construction of dedicated multiscreen cinemas. In January 1974, *Variety* considered the major trends in film in the previous year, pointing to the exhibition sector as experiencing "multiplex fever, on the grounds that two to six screens on one plot have at least the potential for bettering the odds of paying the rent on the single facility."[5] The antecedents of the multiplex were in the double- and triple-screen cinemas located within, or located adjacent to, the shopping malls springing up across the US suburbs in the 1960s. Since the shopping mall and the suburb were the site of subsequent developments and the enlargement of the twins and triplexes to quadruplexes, sixplexes, and eightplexes, the multiplex has come to be defined as a purpose-built cinema of eight or more screens, with a shared lobby and shared projection rooms for multiple screens. More often than not, it was also defined by its adjacent parking and ancillary attractions such as restaurants and arcades. MEDIA Salles sought to define the multiplex as distinct from the multiscreen cinema, with the latter suited to "sites where a traditional theatre has been divided up, and to make specific design a criteria for calling a theatre a multiplex."[6]

Though the criteria for a multiscreen cinema were clear, for the multiplex it was less so, largely as the focus on new, purpose-built multiscreen cinemas could include any number of screens starting from 4–5 and going up to 25 or more. Nonetheless, based upon research of the European market, and echoing the experience of the USA, MEDIA Salles concluded that the term 'multiplex' should apply only to a "purpose-built multi-screen facility of 8 screens or more," with their rationale being that this:

> approach, which aims at evaluating chiefly the efficiency of the multiplex formula (in terms of the degree to which the facilities are utilised), seems preferable to an approach based on the presence of a series of qualitative features (not only the previously mentioned car parks and refreshments, but also screen size, steeply tiered seating, distance between seats, space in the foyers, air conditioning, quality of sound, etc.), which might also be found in theatres which are not multiplexes; moreover these are features that may not be considered equally necessary for the definition of a multiplex as such.[7]

These criteria for the multiplex will largely dictate this study, though it is the case that in many countries official definitions differ. In Japan and South Korea, for instance, the considerable pressures on space and price of real estate, and the necessity of incorporating cinemas into high rise developments, mean that here a multiplex can be smaller. Similarly, in China, the average screens per site are more like five screens. In any event, the key distinction of the multiplex is that, according to Delmestri and Wezel, it represents "a substantive innovation, which aimed at offering a new experience to the visitor."[8]

## STRUCTURE

This book is divided up into four parts: Part I considers the development of the multiplex in the USA, from the 1960s; Part II opens out the discussion of the international rollout starting with the development of the multiplex in Britain from the building of the first purpose-built multiplex cinema in 1985; Parts III and IV consider two large regional case studies, designed to complement the book's previous focus on the USA and Britain and enable a focus on some key territories. These are Europe and Asia-Pacific. Each regional case study has been selected for its own intrinsic value due to the particular political, economic, and/or social/cultural issues raised in particular territories.

Chapter 2 considers the origins of the multiplex in the USA, from the 1960s, and examines the importance of the shopping mall, and the suburb as the main focus for the development of a new kind of cinema—purpose built and multiscreen. In the wake of the *Paramount Decrees* and the relinquishing of their first-run cinemas, the major studios ceded their place to a new group of pan-regional exhibition companies like American Multi-Cinema (AMC), General Cinema Corporation, and National Amusements. The 1960s and early 1970s were a significant period in the development of the multiplex cinema, not simply in terms of the imperatives of design, but in the ways in which cinemas were recast with new forms of marketing, consumerism, leisure, and business management.

Chapter 3 charts the emergence of the multiplex in the USA, with 8, 10, and 12 screens or more, and the ways in which these new forms of cinema signalled a radical change in the perception of where cinemas 'belonged,' as the mall space gave way to the freestanding complex. The diffusion of the multiplex across the country was the result of companies that had previously been regionally based beginning to look further afield, utilising market research in evolving suburbs and forging alliances with large developers. The result was the creation and growth of several national chains (Showcase, United Artists Theatres, Loews, Cineplex Odeon, AMC, Carmike Cinemas, and General Cinema) by the 1980s.

Chapter 4 examines the development of the megaplex—giant multi-screen cinemas with 16–30 screens—in the context of the hegemony of the multiplex as the de facto norm for cinemas in the USA by the 1990s. The megaplex signalled the maturity of the multiscreen cinema and was based upon a new kind of relationship with the major distributors, especially with the development of the film blockbuster.

Chapter 5 considers the decision by a series of US exhibitors to enter the British market, with particular attention to Britain's first multiplex— The Point opened by AMC in 1985 and the first US multiplex opened in a foreign country. As one of the first countries to import the multiplex from the USA, Britain can be considered a kind of 'test bed' for the concept. To this end, some consideration is afforded to the exhibition landscape in the period prior to 1985, especially as 1984 was the nadir of cinema attendance in Britain, when annual admissions were only 54 million.

Chapter 6 considers the complex patterns of ownership and volatile nature of the British market, as the existing exhibition duopoly of Cannon and Odeon was first challenged by US operators such as AMC, Warner Bros., and National Amusements, before this initial wave of US companies either exited the British market or were bought out by domestic and European exhibitors such as UCI, Virgin, and UGC. Chapter 6 also explores the growth of the multiplex and the megaplex cinema in Britain, accounting for a reversal in the downward spiral of cinema attendance as a result of their introduction. Finally, the discussion of ownership is examined through a discussion of the current domination of the market by three exhibitors: Cineworld, Odeon, and Vue.

Chapter 7 considers the location, design, and operation of the multiplex cinema, situating the analysis in the context of urban planning and in particular the easing of planning restrictions in the 1980s and early 1990s, and the development of large out-of-town shopping centres. The emergence of a new kind of out-of-town leisure and shopping culture in the 1980s and 1990s is considered in the context of both economic and political changes, and indicative of a new set of aesthetics around shopping and consumerism. Chapter 7 also contrasts these earlier developments with changes to planning laws and a greater stress on town and city centre regeneration since the late 1990s, which have seen the relocation of the multiplex from the periphery of Britain's towns and cities back to the centre, as the focus of cultural and economic policy has shifted from out-of-town to the city and town centre.

Chapters 8 and 9 consider the ways in which the concept of the multiplex cinema has been embedded in Europe—through both the expansion of US multiplex operators and its adoption by indigenous cinema exhibitors. It considers a series of case studies of territories in which US companies accessed markets more readily—Germany, the Netherlands, and Denmark—and markets where the penetration of US exhibitors was marginal or non-existent—Belgium, France, and the rest of Scandinavia.

Chapters 10 and 11 consider the development of the multiplex in Asia-Pacific, not least because many territories here have witnessed significant growth from a very low base, or are mature markets, in which indigenous operators have dominated both domestically and expanded regionally. Chapter 10 considers Australia, as an example of a mature multiplex market in which domestic exhibitors dominate, having benefitted from greatly increased cinema attendance, and expansion overseas into several global markets. Finally, Chap. 11 examines Japan, South Korea, and China, which are markets that have also expanded rapidly, but which have particular domestic conditions that contrast sharply with each other. South Korea has the world's most enthusiastic cinemagoers per capita, having established the hegemony of the multiplex in 20 years, based entirely on domestic operators. Japan was a multiplex market that was established largely by US and British entrants, though all exited the market subsequently, leaving it to domestic exhibitors. Finally, China is both the fastest growing film market in the world and one of the most complex to do business in and with.

# Notes

1. Larry Gleason, "Multiplex Phenomenon Gains Int'l. Momentum," *The Film Journal* 95, no. 2 (1 July, 1992), 28.
2. Sumner Redstone, *A Passion to Win* (New York: Simon and Schuster, 2012), 78.
3. Ibid., 78.
4. Dodona, *Cinema Industry Research India 2018* (Leicester: Dodona Research, 2018).
5. A. D. Murphy, "Big Film Trend Stories of '73," *Variety* 273, no. 9 (9 January, 1974), 64.
6. Elisabetta Brunella, "Multiplexes in Europe 2000," *MEDIA Salles European Cinema Yearbook 2000.* www.mediasalles.it/yearb00/multiplex.pdf (accessed 11 February, 2019).
7. Ibid., 105–6.
8. Giuseppe Delmestri and Filippo Wezel, "Breaking the Wave: The Contested Legitimation of an Alien Organizational Form," *Journal of International Business Studies* 42 (2011): 828–52, 830.

# PART I

# The Multiplex Cinema in the USA

# CHAPTER 2

# The Origins of the Multiplex in the USA

The USA emerged from the Second World War as the preeminent economic world power, which Miller et al. identified as a "military and diplomatic hegemony articulated to the expansionary needs of its corporations."[1] Of particular importance were Hollywood and both its domestic and overseas market, since unlike its European, Asian, and other overseas rivals, the US film industry continued uninterrupted throughout the war and emerged afterwards with a significant backlog of films available for post-war markets. In the USA, dominance by the major studios was based upon vertical integration, in which they controlled production, distribution, and exhibition, and whilst they theoretically claimed that their markets were open to foreign films, in reality the studios produced enough films to fill the nation's cinemas.[2]

The cultural importance of the cinema to the US population during the Second World War reflected that of countries like Britain, and which was carried over into the peace. The USA had experienced a "golden age" of cinemagoing during the war and in the immediate post-war period, when in 1948 weekly admissions stood at 90 million, which equated to 4.6 billion annually.[3] This was to be a peak as just two years later in 1950, admissions had dropped to 3.01 billion before more than halving to 1.3 billion in 1960 and then dropping to 920 million in 1970.[4] Obtaining numbers of cinemas in the USA, particularly in the period from the end of the Second World War through to the 1970s, is complicated by the presence of drive-in screens and what were often referred to as "indoor" or

© The Author(s) 2019
S. Hanson, *Screening the World*,
https://doi.org/10.1007/978-3-030-18995-2_2

"hardtop" screens (those within cinema buildings). Nevertheless, in 1948 there were 18,631 screens in the USA, of which 17,811 were "indoor," whereupon the number of screens then declined throughout the 1950s and 1960s, though by 1970 it had risen to 13,750 with 10,000 "indoor."[5]

As well as being the zenith of cinemagoing, 1948 also saw the break-up of the structure of major Hollywood studios. A key element of this structure was the extent to which the industry had become vertically integrated, so that the organisation established in one part of the production circuit gained control of the other parts of the production process. To this end, many of the studios came to encompass the spheres of production, distribution, and exhibition, with the box-office revenues providing them with on-going finance. Eight major studios—Paramount Pictures Inc., 20th Century Fox Corporation, RKO, Loew's Incorporated (MGM), Warner Bros. Pictures, Columbia Pictures, Universal Pictures, and United Artists Corporation—all operated first-run cinemas in major cities in order to offer guaranteed outlets for exhibition. In the late 1930s, the federal government began an anti-trust suit against the studios concerned, viewing their dominant control of the exhibition industry as a restraint of trade. In 1948, the US Supreme Court finally outlawed the policy in the case of *U.S. v. Paramount Pictures Inc.*, et al. (known as the *Paramount Decrees*) and the studios were forced to divest themselves of their cinemas.[6] The case was concerned with the vertical integration of the cinema industry and the extent to which this violated the Sherman Act,[7] with the court finding that the studios had "unreasonably restrained trade and commerce in the distribution and exhibition of motion pictures and attempted to monopolize such trade and commerce."[8] However, despite the hope on the part of the US government that the Supreme Court would find vertical integration illegal it did not do so. Rather, it took the view that without what Weisman identified as "conspiratorial design or monopolistic objective," it was "not *ipso facto* illegal."[9] The court ordered the divorcement of distribution and exhibition because it was clear that the studios had monopolised the first-run exhibition of films, denying access to films for independents. In order to prevent future vertical integration, the court would have to prove that a studio was conspiring to restrain trade or seek monopolistic advantage. Ironically, as Holt points out, the decrees explicitly barred only three studios—Warner Bros., 20th Century Fox, and Loew's Incorporated (MGM)—from owning cinemas without court approval.[10] Like RKO (dissolved in 1959), Paramount, whose name is associated with the case, renegotiated with the court and thus was not prevented by the decree from owning cinemas.

Over the following decade, the major Hollywood studios realised that distribution was the key to the film industry and whilst dissolving their interests in exhibition they keenly retained their distribution arms.[11] In overseas territories, the studios established local subsidiaries of home-based distributors, which allowed for the complete control of marketing and promotion, and the retention of profits. Nonetheless, as they surrendered control of their cinemas and increasingly sold them to the new and emerging cinema circuits, many argued that that the loss of this major source of revenue led to cutbacks in the number of films produced by the major studios.[12]

The period after 1948 in the USA was one of enormous economic, social, and cultural change, in which the cinema would have to find a space, literally and metaphorically. The development of multiplex cinemas in the USA ran in tandem with two phenomena—the post-war development of suburbs and the radical transformation in retailing during the 1960s and 1970s. The 1960s and early 1970s were a significant period in the development of the multiplex cinema, not simply in terms of the imperatives of design, but in the ways in which cinemas were recast with new forms of marketing, consumerism, leisure, and business management.

## From Outdoor to Indoor Screens—The New Circuits

In many ways, it is perhaps significant that the major studios *did* lose control of their cinemas, since the development of the multiplex owes much to the dynamism of a group of initially regional exhibitors, often led by charismatic and dominant figureheads such as American Multi-Cinema's (AMC) Stanley Durwood, National Amusements' Sumner Redstone, and General Cinema's Richard Smith, all of whom entered the exhibition business through their family companies. As Melnick and Fuchs observed:

> [t]heir educational backgrounds and postwar ideals enabled them to match their instincts and wits with a more formal business acumen that all three received from the same university: Harvard. Roughly the same age as one another, the three would help refashion the world of motion picture exhibition during its transitional years by mixing the grit and determination of their fathers' generation with the legal and business skills that their higher

education imbued …The showmen of the past were gone forever and a new kind of exhibitor had replaced them: an efficient, postwar entrepreneur operating just outside America's dying downtown and deftly conquering the new American business center: suburbia.[13]

AMC had its origins in 1920 when Edward, Maurice, and Barney Dubinsky, who had performed in tent shows across the US Midwest, bought a Kansas City cinema called the Regent Theater.[14] They began expanding into towns and cities in both Missouri and Kansas and by 1932, their company owned more than 40 cinemas in the two states, though this was scaled back during the Depression. Maurice had died in 1929 and in 1936 Barney retired after suffering a major car accident, leaving Edward in sole control. In 1940, Edward changed his surname from Dubinsky to Durwood, having taken the name from a character—Durwood Belmont— in a 1914 silent film titled *Lena Rivers*.[15] Stanley Durwood, Edward's son, joined the company formally in 1945 having returned from the Second World War, where he had been a navigator for the US Army Air Force. By then the newly renamed Durwood Theaters operated eight cinemas in Kansas and Missouri. For the next five years, the company expanded, with Stanley persuading his father to invest in the burgeoning drive-in sector. However, the late 1940s and early 1950s were also one in which Edward Durwood entered into a series of high-profile lawsuits with Barney (who died in 1948) and his other two brothers—Irwin Dubinsky and H. W. Dubinsky—over ownership and operation of their cinema properties, which was only fully resolved in Edward's favour in 1953, when he was declared sole owner of their theatre properties.[16] In 1960 upon the death of his father, Stanley Durwood assumed control of the company.

The beginnings of General Cinema lay with founder Philip Smith, who purchased the run-down National Theater in Boston in 1922. Smith introduced a series of innovations at the National including a varied weekly programme of films and multiple show times, which, with reduced ticket prices, gave the impetus to what became a small local circuit of 12 cinemas by 1925. The Depression meant contraction and it was not until the 1930s that Smith was able to expand again and this time he did so by capitalising on the growth of the motorcar and developing the drive-in cinema. The company he formed, Mid-West Drive-Ins, operated more than 70 per cent of the drive-ins in the USA by 1940, before diversifying into "hardtops" in 1951 with the opening of what is generally acknowledged as the first shopping centre–based cinema at the Shopper's World Mall in Framingham,

Massachusetts.[17] In 1960, the company became the publicly listed General Drive-In Corp one year before founder Philip Smith died leaving his son, Richard, as President and Chairman of the Board. As if to finally signal the transformation in both the company and the cinemagoing landscape, the company changed its name again to General Cinema Corporation (GCC) in 1964 as Richard Smith undertook to dispose of its drive-in cinemas.

Whilst Sumner Redstone's National Amusements, whose cinemas were branded as Showcase, was considerably smaller than AMC and GCC, its influence in terms of cinema developments, particularly multiplexes, was significant. Sumner Redstone was an extremely powerful figure in the film industry more widely. The company began in Massachusetts in 1936 when Sumner's father Michael Redstone opened the Sunrise Drive-in cinema in Long Island, New York. The company expanded with the popularity of the drive-in and controlled more than 60 sites in the Northeast. Sumner joined the business in 1954. Unlike its major rivals, National Amusements was a private company owned by Redstone, who latterly controlled 80 per cent of the stock and his daughter Shari, who had the remaining 20 per cent.

As beneficiaries of the *Paramount Decrees*, companies like AMC, General Cinema, and National Amusements chose to ignore the city centres and adopted a new-build, suburban-orientated policy of cinema construction. Implicit in this was a new marketing approach based around the multiscreen concept and allied with many of the features of other American leisure forms. Many sites formed part of the burgeoning shopping malls, and the motorcar became a prime consideration.

## The Suburbs, Shopping Malls, and the New Twin Cinemas—'Less Like Theatres and More Like Homes'

The concept of the enclosed, climate-controlled shopping centre was introduced in Minneapolis' Southdale Shopping Centre in 1956, which was promoted as "tomorrow's main street today."[18] Architect Jon Gruen referred to his centre as "introverted," seeing in this a virtue that was absent in the "extroverted" commercial strips that were a characteristic of suburban America.[19] In what came to be called "the malling of America,"[20] it is estimated that some 17,500 were constructed in a 20-year period from 1960 to 1980, whilst by 1990 the total had exceeded 36,000.[21] The malls were built on the edges of metropolitan centres, and the impact on those areas was dramatic, as the mass free parking drew the car-owning suburbanites away from the downtown business and shopping area. Moreover, the malls themselves were different in one other way from the downtown areas and that was in the kinds of consumers they attracted. The downtown business areas, Teaford argued,[22] attracted a diverse range of people whereas the malls reflected the white, middle-class composition of many suburbs.

Stanley Durwood, head of AMC, and one of the cinema circuits that became most identified with the shopping centre, made great play of the company's focus on initial research into potential locations for their cinemas, and particularly demographics, including the income, age, education, and occupation of the surrounding population. Durwood said that he preferred middle-class areas with college-educated, young people since they were "the backbone of the existing motion picture audience, and our future."[23] This research also determined the number of cinemas to be built and the size, based on what it was believed the area could support economically. As cities became ever more identified, in the public consciousness, with decline and poverty, the mall offered, Teaford argued, an "escape from the harsh realities of a heterogeneous metropolis evident in the business district."[24]

Many of the new shopping malls were built to serve existing suburban populations, which were increasingly characterising post-war USA. As Jackson observed:

> [S]uburbia has become the quintessential physical achievement of the United States; it is perhaps more representative of its culture than big cars, tall buildings, or professional football. Suburbia symbolises the fullest, most unadulterated embodiment of contemporary culture; it is a manifestation of such fundamental characteristics of American society as conspicuous consumption, a reliance upon the private automobile, upward mobility, the separation of the family into nuclear units, the widening division between work and leisure, and a tendency towards racial and economic exclusiveness.[25]

The major impetus for the development of the suburbs lay in a series of government policies, especially the Federal Housing Administration (a product of the National Housing Act 1934) and Veterans Administration (created in 1930) via the 1944 Servicemen's Readjustment Act (popularly known as the "GI Bill") which provided low-interest mortgages, with small down payments for more than 11 million new houses, and were largely directed towards the construction of new family homes, away from urban areas.[26] According to the US Census Bureau, the growth of the metropolitan areas in the twentieth century, which went from the minority of the population prior to the Second World War to four out of five people by 2000, was principally accounted for by growth in the suburbs.[27] The movement of people into the suburbs to take up new single-family housing coincided with or was perhaps a product of, a baby boom between 1950 and 1960, in which the population increased by 19 per cent, the highest of any decade except for the 1900–10, a period which was largely characterised by mass immigration.[28] The impact of the suburbs on the US economy was also pronounced since, according to Austin, families there earned incomes "70 per cent higher than those of the rest of the nation" and although "they accounted for less than one-fifth of the total population, suburban families possessed nearly one-third of the spendable income."[29]

The dispersal of more of the population to the suburbs, allied to higher incomes, meant an increased dependence on the motorcar. In their report for the Federal Highways Administration, Lansing and Hendricks calculated that in the suburbs only 4 per cent of all income groups did *not* own a car as opposed to 54 per cent in older central cities.[30] The car ownership rate for all suburbanites was 49 per cent as opposed to 38 per cent in older cities, but households with two or more cars were 47 per cent in the suburbs and only 8 per cent in the older cities.[31] Significantly, the increasing use of cars by women characterised the suburb and this was also important in the development and popularity of the shopping mall. The expansion of the shopping mall in the 1950s and 1960s was also linked inextricably to the development of the new highways being built across the country as a result of the 1956 Federal-Aid Highways Act (known as the National Interstate and Defense Highways Act). Many of the new shopping malls were located at the intersections of major highways. The 1956 act provided federal government support for the construction of more than 40,000 miles of interstate highways, and in particular the building of beltways (by-passes) around major cities. The 1956 act expanded the existing system from 1000 to 41,000 miles and authorised $25 billion to be made available from 1957 to 1969, with the federal government picking up 90 per cent of the cost.[32]

Initially, in the suburbs, Austin argued, the increased time away from work led to the growth of new forms of leisure, of which the cinema was but a minor one.[33] This was due in part to the privatisation of leisure and new home-based entertainment technologies, but also the absence of cinemas in these new areas. However, this would change. The increase in the number of screens throughout the 1960s was largely a result of a sustained boom in the construction of new, initially single-screen but later mainly twin-screen cinemas, in the burgeoning suburbs and in new shopping centres on the periphery of major towns and cities.[34] Writing in 1961, designer A.A. Ostrander observed that "[t]heatres in shopping centers are sure to change the entire entertainment industry in years to come … the people are continuing to move to the suburbs and it's up to the producers and actors to follow them."[35] An editorial in the *Miami Times* argued that "the motion picture industry has finally awakened and is meeting its competition. If the people won't travel far to see a movie, then the movie men meet them more than half way and bring the theatre to their doorstep."[36] In 1962, in one of the first of a series of "Modern Theatre Surveys," *BoxOffice* reflected on the exhibition landscape in 1961 and contended that in terms of "dollars actually spent, there has been nothing like it since 1950."[37] *BoxOffice* noted in 1962 that the major trend emerging was the interest in shopping centres, with 41.2 per cent (70) of new cinemas being located there.[38] On it went, according to *BoxOffice*, with openings in 1963 rising to 240 "four-wall" cinemas, in 1964 to 349, and peaking in 1965 with 352.[39] Whilst construction continued throughout the second half of the 1960s, there was some volatility, blamed in large part on what one exhibitor called "tight money" and the view amongst some exhibitors that the market was oversaturated.[40]

Though it came to be identified with the suburb and the shopping mall, the twin-screen cinema was pioneered in New York City in 1962, when Rugoff Theatres Inc. opened the Cinema I and Cinema II on Third Avenue. The "piggy-back" design, by architect Abraham Geller, placed a 750-seat cinema on top of a 300-seat one below, positioned at 90 degrees to the upper, with both having separate ticket desks, lounges, and waiting areas. It was claimed that the Cinema I and Cinema II were the first cinemas to be built in New York City in a self-contained building since 1932.[41] Rugoff's complex generated considerable interest throughout the industry and was quickly imitated, not least in New York City itself with the opening of Walter Reade's Baronet and Coronet Theatres later in 1962. This was also a "piggy-back" design—what became known as a "cinema-atop-a-cinema"—however unlike Rugoff's new-build, Reade's was a reconstruction of an existing 430-seat cinema called the Baronet, with a new 585-seat cinema added on top—the Coronet. The Baronet had been a notable first-run, arthouse cinema and Reade's proposal was to spend around $1.5 million on architect John J. McNamara's design, which included a joint façade 70 feet high with two box-offices and an escalator to take patrons up to the Coronet.

At this point, the Downtown areas of cities were still seen as important centres for leisure and business, whilst for the cinema circuits the down-town cinema was often seen as the location for longer-run, 'roadshow' features. Throughout the 1950s, Durwood Theaters, in common with other circuits, expanded, purchasing downtown cinemas in Kansas, includ-ing the Liberty Theater and a number that had been owned by some of the major studios including the Empire Theater (RKO), the Paramount Theater (Paramount), and the Midland Theater (Loew's). In addition, it acquired and converted the Ararat Shrine Temple into a cinema in 1959, renaming it the Capri and equipping it to show films first in Todd-AO, and then Cinerama. Within several years, Durwood Theaters owned all of the cinemas in downtown Kansas City, as the company invested heavily in first-run, prestige sites. Stanley Durwood, however, began to sense a change in the economics of exhibition and in particular the development of the suburbs with the concomitant decline in the attraction of the down-town area. In the early 1960s, Ruben Bergendoff of architectural firm Howard Needles Tanmen and Bergendoff took a group of businessmen, with Durwood among them, on a bus tour of the soon-to-open highway system in and around Kansas City. As Mann observed:

> Durwood was impressed. But later, in a private conversation Bergendoff pointed out to Durwood that the highway system ran both ways—if the downtown was allowed to deteriorate, people would quit coming, and the heart of the city would die. From that point on, Durwood pushed to try to revitalize the core of Kansas City. His efforts weren't always successful.[42]

Having assumed control of the company in 1960 on the death of his father, Stanley Durwood undertook a programme of refurbishment of the company's downtown cinemas, including reducing seating numbers and, significantly, the "twinning" of some of these larger cinemas. In 1962, Durwood converted the lounge of the former Loew's Midland Theater, renamed the Saxon, into the 150-seat cinema called the Studio, which had a separate entrance around the corner. Later in the same year, Durwood introduced a 140-seat cinema called the Academy on the mez-zanine floor of the Empire Theater, what Sweeney characterised as a "theatre-within-a-theatre."[43] Like the Studio, the Academy had its own off-street entrance and Durwood envisaged the two theatres-within-a-theatre taking both first-run and also second-run films.

Durwood's sense that the suburbs were becoming more of a focus for shopping and leisure led him to adopt two broad strategies: the first was to try and attract more cinemagoers into the downtown areas and the second was to recognise that the suburbs were a place that exhibitors needed to move into. With regard to the downtown areas, in 1962 Durwood introduced a scheme to reimburse cinemagoers for car parking after 6pm followed by the introduction of a free bus service for any of Durwood Theaters' patrons. Durwood was acutely aware that the substantial investment in the downtown cinemas was a risk but according to Baer, Durwood had faith in the "vehicular mobility of Kansas City," large amounts of parking, and the presence of a major convention centre.[44] Despite the attractions of the suburbs, Durwood believed that he could "bring these people back downtown and firm up their old patterns of shopping and showgoing."[45] What Durwood did not do, however, was to disregard what was happening beyond the downtown area, seeing the suburban development of Kansas City as an additional, rather than alternative, market.

In July 1963, Durwood Theaters opened the Parkway I and II in the high roof of the Promenade at the new Ward Parkway Shopping Center, trumpeted in a publicity flyer as "Kansas City's only climate controlled shopping streets." It consisted of two auditoria side by side—The Parkway I and II seating 400 and 300 respectively—that shared a lobby, ticket office, concession stand, and projection booth, with separate projection and sound equipment. The new twin-cinema, costing approximately $400,000, was of "French Victorian design" with an awning over the entrance that added to the "Parisian atmosphere."[46] Durwood later said that the design of the Parkway twin cinema was serendipitous in that he had originally met with the developer of the Ward Parkway Shopping Center, John Kroh, to discuss the placing of the cinema over a grocery store, only to discover that that the spacing of support columns made it impossible to site a standard-width auditorium. So, Durwood told Kroh, "why not give me two theaters, side by side?"[47]

As the first designated two-screen cinema within a shopping complex, it aroused much industry attention, whilst the shared lobby and single projection booth were also novel. On the opening night, Durwood obtained two prints of *The Great Escape* (1963) and showed it on both screens to an invited audience. At that point, Durwood thought about multiscreen as a way of showing popular films in two spaces, a kind of "overload" system. On that first night, many couples found themselves given tickets for different auditoria because Durwood had not quite considered the seating arrangements.[48] Nonetheless, the concept of showing the same film in different auditoria would, like the design, inform all subsequent multiplex developments. The Ward Parkway development was judged successful enough for Durwood Theaters to embark on a programme of edge-of-downtown and suburban cinema building, which included the Embassy One and Two twin cinemas (300 seats each) in Kansas City's Country Club Plaza in 1964 and the world's first four-screen or quadruplex cinema in Kansas City's Metro Plaza Shopping Center, in 1966. The Embassy twin was located underground in what had been a basement storage area, a site that *BoxOffice* described as a "something out of nothing."[49] The Metro Plaza, in contrast, was located within the shopping centre and had a common box-office around a central lobby and a seven-sided concession stand. Two of the auditoria seated approximately 350, with two smaller ones seating approximately 250 people. The single projection booth allowed for programming flexibility, with films able to be shown on multiple screens and retained for a longer period, perhaps on one screen, if demand was there. Durwood argued that the Metro's four-screen concept had "greater appeal to the entire family, have moderate prices and offer constantly a wide variety of programs."[50]

The success of the Metro Plaza emboldened Durwood to think about larger complexes and also to expand beyond its regional base in the Midwest. In 1969, AMC opened the 1050-seat Fashion Square 4 in La Habra, California, which it claimed was the "world's first free-standing multiplex cinema."[51] There were two auditoriums with 375 seats and two with 200 seats, which shared the same lobby and would, according to the manager, M. Robert Goodfriend, "offer maximum flexibility for booking films which appeal to all ages."[52] When the new cinema was launched on 27 August the newspaper advert offered patrons "4-1 fun."[53] In an article in *BoxOffice*, a leaflet setting forth seven advantages for patrons of multiscreen cinemas, and distributed by AMC at La Habra, was reprinted:

[p]atrons can come to the quad complex every week and see a different picture, as at least one or more theatres will have a change of feature weekly. This, as opposed to a single house playing a long run feature which automatically excludes the area patron having seen the current attraction until the change of program allows another visit to the theatre.[54]

The Fashion Square 4 was the first of several quadruplexes planned by AMC—in Aurora, Colorado; Indian Springs, Kansas City; and Oakland, California. All of them were in new shopping centres and all had a common layout of four auditoria with their own entrances off the common lobby with concession stand, display and toilets, and their own individual marquee.[55] Durwood had also looked to change the name of the company to reflect the reach of their new form of cinema operation, initially settling on American Royal Cinemas (ARC). Though the cinema opened as an ARC, the company's name was soon changed after a successful injunction by the American Royal livestock and horse show in Kansas to cease and desist using the American Royal Cinema name. In 1969, Durwood Theaters became American Multi-Cinema (AMC).

Stanley Durwood had not been the first to recognise the virtues of having more than one auditorium. In Canada in 1948, the eventual co-founder of the giant Cineplex Odeon Corporation, Nathan Taylor, had taken over an unfinished cinema in Elgin, Ottawa, as part of his Twentieth Century Cinemas circuit. The cinema had a 1200-seat auditorium, and on a piece of adjoining land Taylor built an additional 800-seat auditorium, with the intention of showing art films. By the 1950s, Taylor was introducing several innovations at the Elgin, not least of which was the showing of the same film on both screens at times of high demand. Moreover, he also recognised that a film which was initially popular in the larger auditorium could be moved to the smaller one as audiences fell. This ability to retain films for longer was important and heralded what became known as the 'move-over' film or the 'sub-run.'[56]

James Edwards, who ran a large California-based exhibition circuit and who died in 1997, constantly disputed the commonly held view in the industry that Durwood had been responsible for the multiplex. In 1939, Edwards had borrowed $400 to purchase a cinema called the Alhambra on Atlantic Boulevard in Los Angeles and in 1940 he bought the property next door, which he converted into a second auditorium. Though initially the second screen was referred to as the "Annex," with an entrance off the foyer of the Alhambra, it was renamed the Alhambra Twin in the 1970s.[57] Edwards would consistently claim to have been the first to open a multiplex: "[e]veryone since who says they invented [the multiplex] is wrong," he once said.[58]

AMC were not the only company focusing on the new burgeoning malls. By 1970, General Cinema Corp. (GCC) had become the largest operator of shopping centre cinemas in the country,[59] and like its competitors it kept on expanding throughout the 1970s and into the early 1980s. For GCC's Richard Smith, the company's move to the suburbs was in large part due to the vision of his father, who had been "wise enough to see the push towards suburbia and the need for shopping center theatres near people's homes."[60] For exhibitors like GCC the positioning of cinemas in or on the site of the new shopping centres springing up across the USA allowed for the placing of the auditoriums on a single level, with a single projection booth that could serve both auditoriums. Finally, the siting of these new cinemas in or adjacent to shopping centres meant that space was available for large amounts of parking. GCC utilised the same "butterfly" design in developments in Houston's Gulfgate City, Meyerland, and Northline Shopping Centres (all opened in early 1965), and in Philadelphia's North East Shopping Centre (opened in late 1965). Writing in 1965, Richard Smith was of the view that the cinema exhibition industry was living through the "third major change in its basic economic structure in 20 years."[61] The first two changes were television and the rapid growth in drive-in cinemas respectively, whilst the third was the boom in shopping centre cinemas. Barton McLendon, from Republic Theatre Corporation, echoed Smith's optimism when referring to the "Great Return by the theater-going public" spurred on in large part by "the great urban migration" to the "new, spacious and super-convenient shopping centers."[62] Corbett summed up perfectly the development by observing that "the movie theater as technology achieved the ultimate physical *and* symbolic connection to the most obvious form of conspicuous consumption: shopping."[63]

The development of the shopping centre, then, heralded the introduction of new cinemas and circuits that took their aesthetic inspiration from the malls themselves and the surrounding suburbs. Indeed, in 1966, a business writer from the *Dallas Times Herald* suggested that "motion picture theatres are looking less like theatres and more like homes."[64] Drew Eberson (son of noted architect John Eberson), one of the most prolific cinema architects working during this period, outlined his vision of the "theatre of tomorrow" to the 1963 Theatre Owners of America convention, as one "with comfortable seats, plenty of leg room, improved sight lines and a design and interior that will be completely flexible to receive any foreseeable new type of exhibition and projection of pictures and sound."[65] Many of the cinema circuits strived to make the interiors of the new cinemas luxurious, with: "continental"-style seating in the auditoria that was wider and with more space between; soft lighting; spacious and airy lobbies with stone or vinyl walls, and soft furnishings; plush carpeted or marble floors; art works displayed; and large glass frontages designed to offer what was popularly known as an "indoor-outdoor" effect. Eberson's design for the Wayne Theatre, operated by Skouras Theater Corp. in 1964, was described as having an exterior design "in keeping with suburban atmosphere in the shopping center."[66]

However, if the circuits sought to make the interiors luxurious, the architectural design was often compromised by the location, especially for those cinemas within the confines of mall. Invariably the position of the cinema was dictated by the developer, with many located at the end of the mall away from the key "anchor store" in an effort to encourage patrons to venture nearby (what the industry calls footfall). There was often little or no adornment and many cinemas became largely indistinguishable from the shops within the mall itself. As Gomery observed, the dictates of the mall:

> meant stripping all the art deco decoration that had made theatres of the 1930s and 1940s so attractive. Little survived. There was a necessary marquee to announce the films. The lobby consisted of a place to wait, a department store concession stand, and restrooms. The auditorium was a minimalist box with a screen at one end and seats in front.[67]

There was no doubt that the early multiscreen cinemas were often compromised by poor design and building execution. Variously described as "shoe boxes" and "cookie cutter" theatres, with little, if any, architectural merit, the auditoria were often badly soundproofed, had poorly aligned projectors, and were too small. Valentine observed that in one shopping centre in Santa Fe, New Mexico, a "small marquee hidden in a protective corridor … reads simply 'The Movies.'"[68] For the cinema owners, the location within the shopping mall also offered a series of advantages, such as the opportunity to participate in promotional activities and advertising, with little or no cost. However, as Richard J. Wilson, Vice-President of a New England–based cinema circuit, argued, the key ingredients in appealing to cinemagoers were "accessibility to major thoroughfares and interstate highways; ample, adjacent parking; proximity to the so-called anchor-store; and, most emphatically, nearness to main entrance and exit roads."[69] For many exhibitors, Wilson argued, being "relegated to a relatively obscure portion would prove an audience-appeal deterrent."[70] Notwithstanding the importance of position within the burgeoning malls, the economics of the multiscreen cinema had an appeal for exhibitors, based in part on the ability to have two, three, or four screens with a common lobby, concession stand, and the flexibility to operate a staggered time schedule for the showing of films to overcome problems of large numbers of people at the box-office. In responding to the criticism of the new cinemas' design, exhibitor Julian Rifkin argued that:

> [t]hey weren't big arks in which you could feel lonely. We felt very strongly, and I think the customers felt, that there was a small intimate theatre and you had people there and you felt one with the group. So, instead of saying small, boxy, let's say they were intimate.[71]

In an interview in 1989, Stanley Durwood was asked what he thought of his idea to build multiscreen cinemas; he replied that "[i]t was like punching a hole in the floor of your living room and oil coming out. I figured I had about five years to run with the ball before the big guys would overtake me."[72] Though he was not necessarily overtaken, Durwood's innovations and drive to take his regional circuit national were reflected in the emergence of other similar companies, not least GCC and National Amusements. GCC was one of a number of companies that had previously been regionally based but which began to look further afield, utilising market research in developing suburbs and forging alliances with large developers.[73]

Like Stanley Durwood, Sumner Redstone began to recognise that although drive-ins were popular, the future of the cinema lay in "hardtops," especially located in the burgeoning shopping malls. Moreover, Redstone saw the potential in multiscreen cinemas, particularly the twin-cinema, many of which had begun to appear across the USA. In 1964, National Amusements, which was a wholly owned subsidiary of Redstone Theatres, opened their first purpose-built, two-screen cinema in West Springfield, Massachusetts, which was called Cinema I and Cinema II with seating for 1000 and 850 people, and parking for 850 cars. It was the first indoor cinema built in the Western Massachusetts area for 25 years. The design, by the architects William Riseman Associates of Boston, which had two auditoria at right angles from the common lobby area, would be used by National Amusements in its first round of developments in, for example, Toledo, Ohio, and Washington DC, in 1964; and Lawrence and Louisville, Kentucky, in 1965. Unlike Durwood's twin-cinemas and those of other circuits, National Amusements chose not to locate their cinemas within the confines of malls, making them freestanding instead. Moreover, they did not lease space from developers, but preferred to purchase larger tracts of land on which to locate their cinemas along with adjacent free parking. This also meant that National Amusements owned their buildings too, rather than adopting the increasingly common industry approach of selling buildings and leasing them back from developers. According to Shari Redstone, who took over from her father as president in 1999, "the hallmark of the National Amusement business plan is owning the land under out theatres."[74] By the 1970s, the company was also able to capitalise on its freeholding of drive-in sites by replacing them with multiplexes.

The company began to expand quickly throughout the Northeast in the late 1960s and through the 1970s, particularly after Sumner Redstone assumed the role of President and CEO on the retirement of his father in 1967. Using its new Showcase Cinema branding, trademarked in 1965, the company adopted a policy of new-builds and the expansion and development of existing sites. In 1965, the original West Springfield twin was expanded to a triple with the addition of a new auditoria seating over 1000 people, whilst all three screens were equipped with 35mm for CinemaScope and 70mm projectors. The redeveloped West Springfield triple also embodied National Amusements' appeal to what Berggren and Leonard noted was "the average and above average income family," who would be attracted by the location of the cinema "in a suburban location, *away* from shopping centers and industrial parks."[75]

## THE EVE OF THE MULTIPLEX—"THE CINEMA WITH SIX-APPEAL"

In 1969, under its short-lived name Royal American Cinema, Durwood opened the Six West Theatres in the Westroads Shopping Center in Omaha, Nebraska, which the company billed as the "World's First Six-Theatre Complex."[76] As if to demonstrate the potential of the new multi-screen cinema the opening film, *Candy* (1968), was shown on four of the six screens. The 20,000-square-foot complex had two entrances off the shopping centre's Boston Mall, either side of the ticket booth that was located in the mall itself. The six auditoria were placed in a row and seated from 210 to 319 patrons, with three projection booths each serving two screens. The design reflected the pragmatic approach of Durwood, who recognised that the cinemas needed to be standard and that the designs reflect the "pattern of the shopping center's motif."[77] The decision to open in Westroads was a result of prior research by the company which suggested that Omaha could sustain six new cinemas. As Durwood observed: "[w]e planned to build these, then revised our plans. Since all roads converge at the Westroads Shopping Center and since the center is only minutes away from any place in Omaha, we decided to build the six theatres there."[78] Durwood later outlined his rationale for what was beginning to be known as the "new multiplex concept," stating that the provision of four screens and later six "enable us to provide a variety of entertainment in one location. We can present films for children, general audience, and adults, all at the same time."[79] The slogan for the proposed Omni-6 complex in Atlanta was "the cinema with six-appeal."

In 1969, *The Independent Film Journal* reported that the building boom for new cinemas was at a post-war peak, with construction on a "nationwide basis, beyond all predictable levels."[80] In that year, the newly retitled AMC embarked on a two-year, $50 million expansion programme which would see the opening of some 80 multiscreen cinemas across the country.[81] By 1971, AMC had, or were, building 203 screens in 31 cities in 15 states, with plans to build 17 multiscreen cinemas in 13 cities.[82] The 'sixplex' in Omaha was followed within months by others in Dallas' Northtown Mall and the Military Circle Shopping Center in Norfolk, Virginia. The Town and Country 6 in Houston, opened in December 1969, marked a departure by AMC as it was a freestanding complex—"the world's first free-standing six-theatre complex"[83] no less—set apart from the rest of the shopping centre. The layout of the complex, with its six screens ranging in size from 225 to 356 seats and all accessed off the single lobby, was used as the model for a number of subsequent sixplexes in the 1970s and 1980s.[84] Moreover, the sixplex should be considered as the bridge between the first generation of purpose-built, multiscreen cinemas and the multiplexes and megaplexes that would follow.

## NOTES

1. Toby Miller, Nitin Govil, John McMurria, and Richard Maxell, *Global Hollywood* (London: British Film Institute, 2001), 20.
2. See Ian Jarvie, "Free Trade as Cultural Threat: American Film and TV Exports in the Post-War Period," in *Hollywood and Europe: Economics, Culture and National Identity 1945–95*, eds. Geoffrey Nowell-Smith and Steven Ricci (London: British Film Institute, 1998).

3. Bruce Austin, *Immediate Seating: A Look at Movie Audiences* (Belmont: Wadsworth Publishing Inc., 1989), 35.
4. "Cinemas and Their Audiences: Just Holding On," *Screen Digest* (September, 1992), 205.
5. Dodona, *Moviegoing* (Leicester: Dodona Research, 1999), 20.
6. See Ernest Borneman, "United States versus Hollywood: The Case Study of an Antitrust Suit," in *The American Film Industry*, Revised Edition, ed. Tino Balio (Madison: The University of Wisconsin Press, 1985).
7. The Sherman Anti-Trust Act was approved on 2 July, 1890 and was the first Federal Act that outlawed monopolistic business practices.
8. "US District Court Southern District of New York. Text of the Opinion of the Statutory Court in the Antitrust Suit in the Litigation Against Remaining Defendants, 20th Century Fox, Loew's Inc. and Warner Bros.," *BoxOffice* 55, no. 13 (30 July, 1949), 11.
9. Milton C Weisman, "The Paramount Decrees," *The Independent Film Journal* (30 June, 1956), 97.
10. Jennifer Holt, "In Deregulation We Trust: The Synergy of Politics and Industry in Reagan-Era Hollywood," *Film Quarterly* 55, no. 2 (Winter, 2001): 22–9, 24.
11. See Michael Conant, "The Paramount Decrees Reconsidered," *Law and Contemporary Problems* 44, no. 4 (Autumn, 1981): 79–107.
12. See Michelle C. Pautz, "The Decline in Average Weekly Cinema Attendance, 1930–2000," *Issues in Political Economy* 11 (Summer, 2002). http://works.bepress.com/michelle_pautz/11/ (accessed 25 January, 2017).
13. Ross Melnick and Andreas Fuchs, *Cinema Treasures: A New Look at Classic Movie Theaters* (St. Paul: MBI, 2004), 154.
14. See *Kansas City Star*, "In The Days of the Tent Shows: The Life Of The Dubinsky Brothers," (25 February, 1951).
15. "Durwood's the Name!," *Billboard* 52, no. 19 (11 May, 1940), 28.
16. *The Kansas City Star*, "Suit to Durwood: Referee in 20 million-Dollar Action Rules for The Theater Executive," (21 December, 1953).
17. Christopher Grove, "Showbiz Runs in the Family," *Variety* 369, no. 7 (22 December, 1997), 38.
18. Michael Pacione, *Urban Geography: A Global Perspective*, Second Edition (London: Routledge, 2005), 328.
19. Colin Marshall, "Southdale Center: America's First Shopping Mall," *The Guardian*, 6 May, 2015.
20. See William Severini Kowinski, *The Malling of America: An Inside Look at the Great Consumer Paradise* (New York: William Morrow and Co., 1985).
21. Avijit Ghosh and Sara McLafferty, "The Shopping Center: A Restructuring of Post-war Retailing," *Journal of Retailing* 67, no. 3 (Fall, 1991): 253–67, 255.

22. Jon C. Teaford, *The Metropolitan Revolution: The Rise of Post-Urban America* (New York: Columbia University Press, 2006).
23. Stanley Durwood, "The Exhibitors: Show and Teller Time," in *The Movie Business: American Film and Industry Practice*, eds. A. William Bluem and Jason E. Squire (New York: Hastings House, 1972), 220–1.
24. Teaford, *The Metropolitan Revolution*, 89.
25. Kenneth Jackson, *Crabgrass Frontier: The Suburbanisation of the United States* (New York: Oxford University Press, 1985), 4.
26. See Andres Duany, Elizabeth Plater-Zyberk, and Jeff Speck, *Suburban Nation: The Rise of Sprawl and the Decline of the American Dream*, 10th Anniversary Edition (New York: North Point Press, 2010).
27. Frank Hobbs and Nicole Stoops, *Census 2000 Special Reports, Series CENSR-4, Demographic Trends in the 20th Century* (Washington, DC: U.S. Census Bureau, U.S. Government Printing Office, 2002).
28. Ibid., 12.
29. Austin, *Immediate Seating*, 37.
30. John B. Lansing and Garry Hendricks, *Automobile Ownership and Residential Density* (Michigan: Survey Research Center, Institute of Social Research, the University of Ann Arbor, 1967), 16.
31. Ibid., 16.
32. Richard F. Weingroff, *Essential to The National Interest. Public Roads* 69, no. 5 (March/April 2006). https://www.fhwa.dot.gov/publications/publicroads/06mar/07.cfm (accessed 13 February, 2017).
33. Austin, *Immediate Seating*.
34. See Christofer Meissner, "'A Revolutionary Concept in Screen Entertainment': The Emergence of the Twin Movie Theatre, 1962–1964," *Post Script—Essays in Film and the Humanities* 30, no. 3 (Summer, 2011): 64–76.
35. "Theatres in Shopping Centers Can Change the Entertainment Industry," *Back Stage* 2, no. 30 (25 August, 1961), 17.
36. "Awakened Theatre Industry Catching Up With 'Lost' Suburban Patrons," *BoxOffice* 85, no. 11 (6 July, 1964), SE-1.
37. "$54,725,400 Invested in New Theatres in 1961," *BoxOffice* 81, no. 3 (7 May, 1962), 7.
38. "$90,706,500 for 242 New Theatres During '62," *BoxOffice* 82, no. 14 (4 February, 1963), 13.
39. "$97,411,500 Invested in 320 New Theatres," *BoxOffice* 84, no. 16 (10 February 1964), 20; "$131,122,708 in 450 New Theatres, 221 Located in Shopping Centers," *BoxOffice* 86, no. 13 (18 January, 1965), 8; and "$147,836,000 in 454 Theatres, 238 Located in Shopping Centers," *BoxOffice* 88, no. 13 (17 January, 1966), 7.
40. "$120,730,000 Invested in New Theatres," *BoxOffice* 90, no. 13 (16 January, 1967), 23.

41. Al Steen, "Twin Construction May Start a New Trend," *BoxOffice* 81, no. 20 (3 September, 1962), 10.
42. Jennifer Mann, "Obituary of Stanley Durwood," *Kansas City Star*, 16 July, 1999.
43. Marje Sweeney, "New Academy Theatre Joins Galaxy of Durwood Downtown Showcases," *BoxOffice* 88, no. 9 (17 December, 1962), C-1.
44. Joan Baer, "Enlarges Downtown Activity with a Variety of Operations," *BoxOffice* 81, no. 15 (30 July, 1964), 11.
45. Ibid., 11.
46. "Parkway One, Two Adds Paris Touch to Shop Center's Roof Promenade," *BoxOffice* 83, no. 4 (20 May, 1963), C-1.
47. Jay Blickstein, "AMC: First Multiplex: An Accident of Design," *Variety* 350, no. 6 (8 March, 1983), 43.
48. *The Squire*, June, 2005, 27.
49. "Kansas City Twins are Located Below Ground in Former Storage Area," *BoxOffice* 85, no. 26 (19 October, 1964), b10.
50. Quoted in William La Velle, "RAC Opens First Over-and-Under Quad and First Six-Theatre Complex," *BoxOffice* 95, no. (21 April, 1969), a8.
51. "Four-Plex Announced for La Habra, Calif," *BoxOffice* 95, no. 11 (30 June, 1969), W6.
52. Ibid.
53. See http://cinematreasures.org/theaters/13539 (accessed 23 July, 2017).
54. "Will Automation Revolutionize the Industry?," *BoxOffice* 95, no. 26 (13 October, 1969), a9.
55. See "Construction News: Four AMC Quadruple Complexes Announced," *The Independent Film Journal* 66, no. 8 (16 September, 1970), 4.
56. Philip Turner, *Cineplex Odeon: An Outline History* (St. Paul's Cray: Brantwood Books, 1998), 2.
57. See Cinema Treasures. http://cinematreasures.org/theaters/9364 (accessed 23 April, 2018).
58. Greg Johnson, Peter Noah, and Scott Martelle, "O.C. Theater Magnate James Edwards Sr. Dies," *LA Times*, 27 April 1997. http://articles.latimes.com/1997-04-27/news/mn-53092_1_james-edwards-sr/2 (accessed 1 November, 2018).
59. Grove. "Showbiz runs in the family," 38.
60. "More Harvard Than Hollywood: A Look at GCC's Richard A. Smith," *BoxOffice* 106, no. 24 (24 March, 1975), NE4.
61. Richard A. Smith, "Abundant Potential for Progress in Shopping Center Theatres," *BoxOffice* 87, no. 12 (12 July, 1965), 60.
62. Barton R. McLendon, "Experience of the Recent Past Charts Course for Next Decade," *BoxOffice* 87, no. 12 (12 July, 1965), 61.

63. Kevin J. Corbett, "The Big Picture: Theatrical Moviegoing, Digital Television, and Beyond the Substitution Effect," *Cinema Journal* 40, no. 2 (Winter, 2001): 17–34, 26.
64. Jim Kendall, writing in the *Dallas Times Herald*, 10 July, 1966, quoted in "Exhibition's First Building Boom in 30 Years Marked by New Designs," 89 (18 July, 1966), SW-5.
65. "Drew Eberson Describes 'Theatre of Tomorrow'," *BoxOffice* 84, no. 2 (4 November, 1963), 14.
66. Wesley Trout, "Lively Interiors Enhance New Theatre," *BoxOffice* 85, no. 16 (10 August, 1964), a22.
67. Douglas Gomery, *Shared Pleasures: A History of Movie Presentation in the United States* (London: British Film Institute, 1992), 95.
68. Maggie Valentine, *The Show Starts on the Sidewalk: An Architectural History of the Movie Theatre, Starring S. Charles Lee* (New Haven and London: Yale University Press, 1994), 182.
69. Allen M. Widem, "New Shopping Center, Theatre Construction Trends Toward Expansion on Existing Sites," *BoxOffice* 109, no. 17 (2 August, 1976), 11.
70. Ibid., 12.
71. Quoted in Barbara Stones, *America Goes to the Movies: 100 Years of Motion Picture Exhibition* (Hollywood: National Association of Theater Owners, 1993), 222.
72. Tom Matthews, "Stanley H. Durwood: The Man Who Invented the Multiplex," *BoxOffice* 125, no. 10 (1 October, 1989), 58.
73. See Stones, *America Goes to the Movies*.
74. Shari Redstone, "The Exhibition Business," in *The Movie Business Book*. Third Edition, ed. Jason E. Squire (New York: Foreside, 2004), 388.
75. G. M. Berggren, and K. R. Leonard, "Cinema 3 Opening an Historic Event," *BoxOffice* 91, no. 13 (17 July, 1965), a6.
76. "World's First Six-Theatre Complex Opens in Omaha Under ARC Banner," *BoxOffice* 94, no. 15 (27 January, 1969), NC1.
77. Durwood, "The Exhibitors: Show and Teller Time," 221.
78. "Kansas City Durwood Circuit to Build Six-Theatre Unit in Omaha Center," *BoxOffice* 92, no. 5 (20 November, 1967), 6.
79. "AMC Quad Announced for Harrisburg, Pa.," *BoxOffice* 102, nos. 10–11 (18 December, 1972), E-9.
80. "Mini-Multi Boom on: See Post-War Theatre Building Peak in 1969," *The Independent Film Journal* 63, no. 13 (26 May, 1969), 9.
81. "Durwood Plans Three Quads in One Complex," *The Independent Film Journal* 64, no. 9 (30 September, 1969), 15a.
82. "AMC in Huge Expansion Plan for 1971 With 70 Auditoriums in 13 Cities," *BoxOffice* 98, no. 26 (12 April, 1971), 4.

83. "RAC Breaks Ground for Houston Six," *BoxOffice* 94, no. 26 (14 April, 1969), SW3.

84. Christofer Meissner, *Six Screens for Suburbia: The Rise of the Multiplex Movie Theatre in Kansas City and the Transformation of American Film Exhibition, 1963–1980.* (Unpublished PhD thesis submitted to the Department of Theatre and Film and the Faculty of the Graduate School of the University of Kansas, 2004), 190.

# CHAPTER 3

# The Emergence of the Multiplex in the USA

With their antecedents in the twin-screen cinemas of the early 1960s shopping mall and the freestanding 'quadruplexes' and 'sixplexes' of the late 1960s, the new multiplex cinemas that emerged in the 1970s began to get larger and larger. As the exhibition industry entered the 1980s, the newer cinemas were growing in size, and there was a trend for larger complexes featuring eight, ten, or more auditoria, coupled with a realisation that the cinemas had outgrown the shopping mall. What emerged was the freestanding cinema complex, still commonly, but not always, sited adjacent to shopping malls, though as Paul observed many of these new multiplexes might be best described as the "theater *as* mall," in that many were built on new sites with easy access to highways.[1] After 1980, the steady decline in drive-in screens was offset by a growing number of 'indoor' screens, which in 1985 surpassed the total in 1948. By the end of the 1980s, there existed more screens than at any time in history—although, they were primarily in multiplexes and not traditional single-screen cinemas.[2] Audiences too were growing with annual admissions passing one billion in 1980.[3]

## "Twinning is Good, Multiplexing is Better"

Stanley Durwood argued in 1971, when announcing that American Multi-Cinema (AMC) would build 17 multiscreen cinemas in 13 cities, that in pioneering the multiscreen concept AMC had "started by throwing 'the book' away, and with it all the outdated practices of our industry," adding

© The Author(s) 2019
S. Hanson, *Screening the World*,
https://doi.org/10.1007/978-3-030-18995-2_3

that "[w]e feel we are the showmen of today. We know we have to be the showmen of tomorrow."[4] For Durwood, the expansion would go on throughout the 1970s, with AMC focussing almost exclusively on out-of-town, shopping centre sites, so much so that for many shoppers across the country AMC would become as familiar a presence as Sears, JCPenney, and Woolworths.[5] AMC symbolised a new, aggressive form of cinema operation, taking advantage not only of relatively low-cost sites and buildings, but also of new forms of employment and managerial practices. As Gomery pointed out, a site with six screens required only one concession counter staffed by part-time students, one projectionist, and one manager who also sold the tickets, which allowed AMC to spread the running costs of the cinema over the revenue base.[6] The multiscreens also enabled AMC to increase their market reach by offering a range of films to a wide audience demographic and capitalise on popular films by extending runs and showing on more than one screen.

By the mid-1970s, it was estimated that there were 127 four-screen, 6 five-screen, 17 six-screen, and 4 seven-screen complexes in the USA,[7] with many circuits acknowledging that the days of the twin cinema were numbered and that the future was larger four-, five-, and six-screen complexes with auditoria seating 350 to 500. As Bud Cadiff, owner of MBC Construction Co., put it, "twinning is good, multiplexing is better."[8] Indeed, the average cinema had declined in size from 750 seats in 1950 to 500 seats in 1977.[9] Bob Beacher, from cinema builder Forest Bay Construction Corp., observed that:

[i]n a lot of ways, the twin theater is obsolete because you need the small auditoriums so you can put in more product. Your [sic] better off completely selling out your auditorium and picking up an overflow into your satellite auditoriums. The people who want to see the picture are probably going to come back anyway, and in the meantime they saw another picture and they stopped at your concession stand.[10]

As one exhibitor pointed out, the multiplexes that emerged in the 1980s and 1990s owed a debt to the innovations of the 1960s and the twin cinemas; "the neat thing about the six-plex and the five-plex and the triple was that they were *designed* to be multiplex theatres. We used the information that was available in those days to provide customers with a much more pleasing atmosphere."[11]

Having opened the first sixplex in 1969, AMC completed what was believed to be the first seven-screen complex in 1972 at Southwyck Mall in south Toledo, Ohio. The Southwyck Seven, as it was called, covered 20,000 square feet with 2000 seats in total, whilst the auditoria accommodated from 150 to 350 cinemagoers. The complex had the by-now-common single lobby, box-office, and concession stand but also included an 84-seat fast-food restaurant. AMC's George Kieffer set out the rationale for the complex in the context of largely empty older single-screen cinemas:

> [t]he vast 3,000-seat downtown film house just doesn't work anymore, so we developed the cluster of small theatres which only take a few hundred people to fill. The psychological impact of this, as compared with the same number of patrons in a huge theatre, can't be underrated.[12]

The complex at Southwyck would be one of many examples whereby screens were added subsequently as demand increased and space became available. In 1976, the complex became the Southwyck Eight when an additional 170-seat auditorium was built in the space formally occupied by the concession stand and fast-food facility, necessitating the construction of a new stand. AMC added a further three screens in 1985, though these had a separate entrance onto the street, and aimed to programme art films in two of the new screens, in an effort to encourage new audiences who would otherwise have to travel to Detroit or Ann Arbor. Jim Bond, the cinema's manager, set out the hard-nosed business case for the plans: "[w]e are a business, not the Toledo Museum of Art. If the art and foreign films don't draw then we'll drop them."[13]

The opening of ever-bigger complexes was not just the domain of the major exhibitors though. Cobb Theatres opened the Cinema City 8 in the Roebuck Shopping City, East Birmingham, Alabama, in June 1978. There was a single lobby with a box-office and two concession stands, whilst the eight auditoria had between 250 and 300 seats, and were fully automated, with only two projectionists needed for all eight screens. Eight Christie consoles were installed with Christie Autowind III projectors and platters, which could accommodate four and a half hours of programming with no necessity to rewind, since the projectionists simply rethreaded the film through the projector for the next showing. This, Christie claimed, reduced manpower and other costs.[14] Another significant innovation, that would presage later developments, was the 15,000-square-foot "Super

Cellar" beneath the auditoria, which housed: a restaurant selling ice cream soda, hamburgers, hot dogs, and pizza; a movie boutique selling records, T-shirts, posters, and books; and a 2000-square-foot games room with pinball and games machines.[15] Cobb Theatres were based in Alabama and were one of the first circuits in the south, having been founded in 1946. They grew rapidly in the 1970s and 1980s as a result of acquisition, though they were themselves merged with Regal Cinemas in 1997.

In April 1979, Canadian exhibitor Cineplex Odeon opened an 18-screen cinema, some years before other circuits opened complexes with ten or more screens. The development in Toronto's Eaton Centre was an ambitious scheme, which cost over C$2 million. Operating over two levels, with mezzanine floors, 18 separate auditoria, and with significant space for car parking, it was Canada's first true multiplex.[16] Cineplex Odeon's flamboyant head was Garth Drabinsky, who in 1979 had persuaded Nathan A. Taylor, who had been involved in the Canadian exhibition industry since working in his father's Monarch cinema circuit in 1922, to invest C$1 million in the Eaton Centre scheme. Drabinsky coined the term "Cineplex"—a conflation of cinema and complex—for the new site when it opened. It was immediately branded, Linck argued, "as the Titanic of exhibition by doubters."[17] The layout of the 40,000-square-foot complex did, however, force a number of compromises such as the small size of the auditoria, the largest of which accommodated 160 people and the smallest only 40. In addition, the lack of space dictated the use of 16mm projectors rather than the bulkier 35mm ones, though the small auditoria did not demand much of a throw. For these reasons and the initial lack of access to many first-run feature films, the two major circuits in Canada—Famous Players and Canadian Odeon Theatres—did not take the 18-plex seriously, predicting that the company would go out of business within a year. However, the cinema prospered, as weekly takings exceeded C$50,000 for the first three years.[18] Part of the success of Cineplex was their formula for programming, in which popular films were shown on up to four screens, which decreased in number as the film's popularity waned. This meant that no one had to be turned away. Moreover, the lack of access to many first-run films meant that in addition to second-run features the Cineplex showed a range of foreign-language and independent films. As architect Mandel Sprachman argued, "there isn't a hope in hell of filling a 2,000-seat theater anymore ... Being one of 25 people watching a movie in a 2,000-seat house is not a pleasant experience. That situation will never arise at Cineplex."[19]

The success of the Eaton Centre Cineplex drew the company to the attention of other shopping centre developers and within two years Cineplex had 111 screens across a number of other sites in the Toronto area as well as Vancouver and Calgary, with plans to expand in several other Canadian cities. In 1984, Cineplex purchased the rival Canadian Odeon Theatre circuit in what *Variety* described as a case of the "Mighty mouse swallowing an elephant."[20] A year later, the renamed Cineplex Odeon Corp. purchased the US-based Plitt Theatres circuit seeking to consolidate its position in the US market, whereupon it undertook an aggressive acquisition campaign in 1986–87 buying up smaller regional circuits including RKO Century, Warner Theaters, and Walter Reade.

In 1979, a regional circuit, Cinema Centers Corp. from Boston, had opened an eight-screen complex in Holyoke Mall in Springfield, Massachusetts, and then a ten-screen multiplex in Crossgates Mall in Albany, New York, in 1984. Though not as large as Toronto's Cineplex, Crossgates was the more significant development in terms of its heralding of a trend towards the widespread diffusion of the archetypal 10- to 12-screen multiscreen cinema. In the first instance the mall itself was one of many being built in large-population centres on major highways—in the case of Crossgates at the intersection of the Northway and Thruway interstates within an area of "proven filmgoing populations" near the State University of New York and the State Office Campus.[21] At more than one million square feet the mall incorporated four department stores and more than 30 stores, ranged across three tiers, with the Crossgates tenplex located on its own mezzanine level, and accommodating 3000 people in auditoria ranging from 200 to 500 seats. There were two ticket kiosks utilising the latest computerised, pre-booking systems and two concession stands, with the mall corridors providing what was in effect an extended lobby.[22] Cinema Centers Corp.'s vice-president Malcolm Green said that since there were ten screens "we will be able to offer most first run features an auditorium on national availability at every season," adding that "an expanded run of two to four or more area screens will be in the best interest of all parties, as it is in every important market."[23]

As the 1980s dawned, the industry was acknowledging that the minimum number of screens required to make a complex viable was six but with eight increasingly common. The mood of optimism, though tempered by the impending threat of financial downturn in the US economy, was bolstered by the Motion Picture Association of America's (MPAA) figures for cinema attendance by those aged 12 and over, which had increased 20.4 per cent between 1969 and 1979 to reach 113.7 million visits.[24] Moreover, between 1970 and 1980 the number of screens in the USA increased from 10,000 to 14,171, with almost all of the increases in multiscreen cinemas.[25] This growth in multiscreen cinemas was in large part the result of two developments: the requirement for economies of scale on the part of the new national circuits and the changing practices of film distributors. In the case of the former, circuits like AMC and National Amusements increasingly turned away from smaller auditoria, stating that 400–500 seats were the minimum, even with the emerging eight- and ten-screen multiplexes. The desire for larger auditoria was also dictated by the necessity of generating enough ticket sales to satisfy the distributors' demand for "larger gross potential."[26] Companies like AMC began to argue that increased numbers of auditoria (at least eight) meant greater flexibility and efficiency in booking films. Moreover, exclusive access to films and licencing was disappearing, whilst there was a trend in the 1980s for many states to outlaw practices like blind bidding. This meant that distributors looked increasingly to as wide a release pattern as possible, supported by national advertising, so that all cinemas became first-run cinemas. As Paul argued, their status as first-run cinemas, rather than their architecture, assigned them quality, as had been the case with the old downtown picture palaces.[27]

## Distribution and Programming—"Multiplexes are Hungry Monsters"

The consolidation of the multiplex as the preeminent cinema form in the 1980s ran concurrently with a series of trends in the production, distribution, and exhibition of films in the USA. In the 1980s, there was a dramatic growth in both the number of films made and total box-office receipts, so that by 1990, 410 films were rated and released in the USA as opposed to 233 in 1980.[28] In the same period the total box-office rose from $2.75 billion in 1980 to $5.02 billion in 1990.[29] If one includes the 1990s, when the exhibition industry began to introduce larger multiplexes, then the box-office continued to increase until it reached $6.91 billion by 2000.[30] Throughout the 1980s and 1990s, the tendency on the part of the major studios was for a smaller number of blockbuster films to dominate their release schedules. These blockbusters were characterised in part by: their high production and marketing budgets, the latter of which could equal the former; their financial importance to the studios making them; and their appeal to as wide an audience as possible. Jancovich and Faire argued that "the term 'blockbuster' has come to be associated with the multiplex, and the two are intimately related."[31] To this end they were often referred to as "tent pole" features since their profits are able to support the entire studio.[32] Increasingly, the studios looked to sequels and the summer period was of particular importance, when they released many of these tent pole features, allied to the fetishisation of the opening three-day weekend. In the 13-week summer season of 1989, for instance, the US box-office was estimated to be $1.8 billion, which would have accounted for 40 per cent of the total annual gross in 25 per cent of the playing time; whilst Tim Burton's *Batman* took approximately $42 million in its first three days.[33]

For distributors, the strategy was increasingly to opt for a saturation release in which the film was booked into the largest number of cinemas, particularly in the opening weekend. As Batey argued, multiplexes were "hungry monsters."[34] To this end, the multiplex cinema offered a range of possibilities for managing a film during its theatrical release, before it moved into the new home video market, some six months later. Upon opening, the multiplex could show the film on one or more screens at a wide range of starting times, often utilising multiple screens at the same time. This saturation opening, often on multiple screens, added to the mystique and importance of the opening weekend in terms of box-office. Thereafter, as the film's popularity waned it could be moved onto smaller screens, still maintaining its visibility. Though blockbusters, and the importance of the opening weekend, had begun to appear from the mid-1970s onwards and thus predated the development of the multiplex, it is possible that distributors learnt to use the multiplex to suit their release strategies and manage the movie during its release.[35]

The focus on opening weekends and the small number of tent-pole features that did extremely well at the box-office acted to disguise one significant trend during the 1980s and 1990s and that was the stagnation in the box-office earnings per screen. This was less a function of the numbers of people going to the cinema than the proliferation of multiplexes and the corresponding number of smaller screens with smaller seating capacities. For instance, in the 15 years after 1983, the average return per screen of $200,000 did not change, despite a five-fold increase in the cost to the studios of producing and distributing films.[36] This decline was arrested subsequently—$261,185 in 2000—largely as a result of increases in admissions prices, which in 1999 passed the symbolic $5 average mark.[37] Indeed, by 2017 the lowest cinema attendance this century was offset by increases in ticket prices that were an average of $8.65, which meant that box-office was substantially higher and had been rising year on year.[38] Despite a continued increase in the number of screens added year-on-year, annual admissions per screen also continued to fall during the 1990s from 50,000 per screen in 1990 to 40,000 in 2000—the lowest in history up to that point.[39] Despite some rallying (by 2003 it was back to the figure of 1992) the trend has been resolutely downwards so that in 2017 admissions per screen were 37,896.[40]

## DESIGN OF THE MULTIPLEX—"THE JAPANESE CAR OF MOVIE THEATRES"

As we have seen, the development of the multiplex cinema ran in tandem with the development of the shopping mall, opening first within the confines of the mall and then emerging as freestanding structures in their own right. The major imperatives in the design of the multiplex were largely the same as those which concerned the developers of the first twin cinemas in the 1960s: that they should be simple designs, have comfortable environments, and promote ticket sales and concession spend with the lowest overhead costs.[41] As the twins became fourplexes, sixplexes, eightplexes, and onwards, then the increasingly freestanding structures sought to establish a new design aesthetic. However, those sites were very often simply large, square metal and concrete boxes, the epitome of Venturi, Brown, and Izenour's "Ducks" and "Decorated Sheds,"[42] in which ornament was present only on the façade, signifying the iconography of the building as a cinema. Arguing in 1987, that the art and science of multiplex design was in its infancy and only then being taken seriously, architect Girard Kinney observed that the:

> major design decisions which determine the overriding form and configuration of the building and decisions which identity the majority of the materials to be used to construct the building, should be driven not by "aesthetic" considerations but rather, by the activities which are intended to occur in and around the building.[43]

Initially, the building design had to take its inspiration from the malls in which the multiplexes resided or were close by. Whilst for the large circuits this could be a constraint it was often an opportunity to standardise designs and benefit from what Paul called the "cost-effectiveness of uniformity ... common and infinitely repeatable architectural design, somewhat modular in approach so that the number of screens actually contained within... won't change the overall look of the theatre."[44] In the interior, the aesthetics of the buildings took their cues from a series of other sites of leisure and consumption, notably the fast-food restaurant. They were increasingly what Edgerton characterised as examples of an "evolving language of American commercialese," which communicated: "bright colors, vinyl and laminated covered furnishings, plastic panel art, sloping wall and ceiling design and glaring lights."[45] In part, this was a response to the demand from distributors, in the face of the growth in the home video market, to make the cinema "more of an event" and "bright and cheerful," most commonly via larger lobby areas, better seats, and a more attractive interior design.[46] They aimed, added Paul, to "posit an audience familiar and comfortable with mass retailing strategies."[47] Multiplexes were a clue, therefore, to a process which the American sociologist Ritzer called McDonalidization "by which the principles of the fast-food restaurant are coming to dominate more and more sectors of American society as well as of the rest of the world."[48] McDonaldization is characteristic of a particular form of capitalist efficiency, in which a large transnational business organisation is run with the smallest possible central organisation and it has been extended beyond the sale of food to encompass other forms of goods and services. It is articulated through the four principles of efficiency, calculability, predictability, and control. The proliferation of multiplexes in the USA subsequently, seemed to mirror the adoption and development of these new forms of business and cultural practices.[49]

Matthews was unequivocal in his reading of the design of the multiplex as largely dictated by pragmatism:

[t]he multiplex, while increasingly elaborate, remains restricted by space and the almighty dollar, with every single square foot put to its best practical use. The multiplex has become the Japanese car of movie theatres—compact, efficient and often lacking in individual style.[50]

This reflected the intense competition that existed between exhibitors. As Wayne Kullander of Harkins Theaters pointed out, "our aim is to increase the frequency of moviegoing—and within a reasonable budget."[51] Nonetheless, the tension lay in establishing an identity for the new buildings that, according to architect Paul Ladensack, communicated a "sense of place and magic," achieved, he felt, by a "modern design that would elevate the experience of cinema and make it playful at the same time."[52] As the large freestanding multiplexes but more particularly mega-plexes emerged, this "playful" element found expression in a series of post-modern flourishes and motifs, with an emphasis often on spectacle, eclecticism, and pastiche. Though many of the first wave of multiplexes were essentially metal boxes, Heathcote argues there emerged another approach and that was a focus on a cinema as "environment," a space of corporate recognition and recognisable imagery.[53]

Architects like Jon Verde, who designed AMC's 30-screen, Block 30 in Cedar Rapids, Michigan, epitomised this new form of design, seeking a "paradoxical mix of commercialism and art, of in-your-face franchising and an architecture which is genuinely good fun."[54] Overwhelmingly, though, the aesthetics of the multiplex reflected the contemporary impor-tance of consumption as the prime determinant of the economy, and this found its apotheosis in the enlarged lobby area. Here was the audience's first point of contact with the multiplex, in which the identity and ethos of the cinema were firmly established. This housed the box-office and the concession counter, and served as a giant advertising and promotional zone for present and forthcoming films; which one exhibitor characterised as both "essentially social spaces" and "the theater's profit centre."[55] In a highly critical reading of the contemporary cinema space Kupfer argued that it was not possible to see the multiplex as a cinema at all:

> [t]he multiplex movie building cannot really be called a theatre. It is just another business, with no special decor or architectural character. It is a device designed to deliver a commodity effortlessly. The building's exterior does little to reveal its cinematic purpose. Often embedded in a series of similar structures, the multiplex cinema looks like a small office building. We are not in a separate building, marking off a separate space. The screen and the viewing room itself are usually cut down, mini-sized. The movie rooms are packaged two, four, six, even twelve to a slab. The result is that the sym-bolic, celebratory nature of the theatre is destroyed as movie-viewing is reduced to another commodity.[56]

Aside from branding, it was possible to see many multiplexes and mega-plexes as broadly similar in design; indeed, as Regal Cinema's Keith Thompson admitted, "[o]ur theatres don't differ dramatically from other megaplexes being built. Our industry is good at copying one another."[57] As each circuit introduced its innovations and emergent technologies— Dolby, THX, and Sony Dynamic Digital Surround sound; seats with cup-holders, stadium seating and reclining seats; advance credit card booking; and ever more elaborate concession offers—then these became widely adopted by competitors.

Many exhibitors were also increasingly citing the concerns around the burgeoning video and cable TV market and, in an echo of the debates around television and new screen technologies in the 1950s, saw bigger auditoria and importantly bigger screens as a way of distinguishing cinema. National Amusement's Sumner Redstone warned in 1982 of the "possible catastrophic competition from home entertainment technology" and argued that "building small theatres is not the way to offer people the luxury required to attract them."[58]

Notwithstanding this, as the twin cinema emerged in the 1960s, the multiplex in the 1980s, and the megaplex, which emerged in the 1990s, the one consistent feature was the diminution in the number of seats in the auditoria. A giant complex like the Regal Hollywood 27 in Nashville, opened in 1998, had over 4800 seats but the largest auditoria contained just 431 seats, whilst 14 of the screens seated only 107 people. For larger multiplexes, the limit seemed to be 500 seats, which would be considered a large audience. Of course, the development of interlocking and newer digital technologies has meant that exhibitors have the ability of showing one print on more than one screen in response to demand.

## CREATION OF THE NEW "GIANTS OF EXHIBITION"

Richard Smith's General Cinema Corporation (GCC), Stanley Durwood's AMC, and Garth Drabinsky's Cineplex Odeon had emerged, along with Carmike Cinemas and United Artists Theatres, in the 1980s as the 'Big Five' of national circuits; which had all expanded through a building boom in the 1960s and 1970s, and which in terms of size and power came to replace the cinema circuits once owned and run by the 'Big Five' major Hollywood studios. In addition, there were a series of other circuits, all of which were in the process of expanding beyond their regional strongholds or were doing so as a result of merger and acquisition—Loews Theaters, Cinemark, and Sumner's National Amusements. By the end of the 1980s, the 'Big Five,' or the new "Giants of Exhibition,"[59] controlled some 30 per cent of cinema screens in the USA.[60] As the decade progressed these large national circuits began to open ever-larger complexes and swallow up smaller circuits in a process of merger and acquisition.

By the mid-1980s, GCC had become the largest circuit in the USA in terms of screens and claimed to be the largest in terms of box-office revenues and attendances. During this period GCC had opened cinemas in more than 30 states but the building in the later 1980s and early 1990s, particularly of the new multiplex cinemas, was focused primarily on what GCC termed the "nation's sunbelt"—from the Carolinas on the East Coast to San Francisco on the West Coast.[61] As two of the fastest growing states in terms of population and economic growth, GCC, like their competitors, looked more specifically to Texas and Florida, and such was their profitability that the expansion could be funded without recourse to major borrowing. In addition to announcing new builds every year it also recognised the importance of modernising its older cinemas; announcing in 1984 a "four-year masterplan" to renovate 75 per cent of its cinemas under the auspices of a "theatre design committee" drawn from various sectors of the company.[62]

Cineplex Odeon's Garth Drabinsky had sought to get into the US cinema market almost from the beginning, and when they did it was with the ambitious 14-screen multiplex in the Los Angeles Beverly Center, opened to the public on 16 July 1982—"Cineplex Day."[63] At that time the largest cinema in the USA, the 25,000-square-foot complex with its 14 screens was located on the eighth floor of the huge shopping complex, though its total seating capacity of 1650 seats was modest given the size of the site. Nevertheless, the Beverley 14, as it was known, had some key innovations including a single projection box for all 14 screens and a mechanical 'interlock' which allowed for a single print to be shown on up to six screens simultaneously.[64] A computerised booking system was installed that allowed cinemagoers to purchase tickets in advance for same-day screenings, whilst lifts and elevators helped access the mass free parking incorporated into the shopping centre's street and four parking levels.[65] Signalling a proposed "invasion" of the USA, Cineplex Odeon also set out proposals for similar multiplexes in Houston and Denver, amongst others, though the reality was that having experienced a series of financial debt problems in the period after opening the Beverly 14, the cinema was sold to Taubman Theatres, a subsidiary of Taubman Co., the shopping mall's developer, for $3.5 million on a buy-back basis at a later date.[66] This kind of real estate deal would become more common as the century progressed.

Ten years after opening their first cinema in the Eaton Centre, Cineplex Odeon became the second largest circuit in North America with over 1900 screens in 497 sites, whilst more than two-thirds of its sites were located across 20 US states.[67] However, the expansion, along with the efforts to forge a vertically integrated company, meant that the levels of debt being built up were considerable. By 1989, Cineplex Odeon owed more than $700 million to 19 banks, whilst its proposed $500 million theme park in Orlando—a joint development with MCA Inc.—was late and over budget.[68] After a period of internal conflict, which in part rested on his management style, Drabinsky was forced out of the company in December 1989. The new management team, led by Allen Karp, began restructuring the debt and looking more methodically at proposed sites and projects, whilst seeking to increase revenue per patron.

These kinds of new and larger multiplexes meant that what distinguished several of the circuits were their average screens per site. As the pioneer of the 6- and 7-screen cinema, AMC began to build 8- to 12-screen complexes during the accelerated expansion of the 1980s, funded in part by the floating of the newly renamed AMC Entertainment Inc. in 1983 and the sale of 12 per cent of the company's stock. In October 1987, AMC opened its flagship Century 14, which was the company's most expensive site—costing in the region of $5 million—and covering 68,000 square feet of the enlarged Century City Shopping Centre, on the former backlot of 20th Century Studios in Los Angeles. All of the projection systems were automated, whilst the computerised box-office system had the facility to handle advance ticket sales. The auditoria seated from 200 to 550 people, with the five largest equipped with 70mm and THX stereo sound systems, whilst the dedicated car park had 3500 spaces. Matthews felt that the complex was "breath-taking" and "makes one wonder if multiplex palaces like this are the wave of the future."[69] The complex incorporated a "theatre within a theatre" with two separate concession areas selling traditional items and another a "more eclectic menu with the speciality filmgoer in mind."[70] The speciality filmgoer would be catered for by the provision for "art films," which was less about building an audience for these films as much as taking the existing audience from rival Cineplex's four-screen Century Plaza over the road and Cineplex's 13-screen multiplex in the nearby Beverly Center, with AMC's schedule matching day and date of these existing cinemas.[71] AMC's Greg Rutkowski argued that "the thing that makes Century City attractive to AMC are the two D's: Density and Demographics. Century City is the cradle of film activity on the West Coast, and AMC needed a visible position in the market."[72]

AMC had an important advantage over several other companies in that their circuit was modern, with a third of all of its screens in the late 1980s less than three years old, whilst their sites had an average of 5.8 screens per site compared to 3.7, which was the national average.[73] However, like many other circuits, the expansion was also fuelled by the taking on of considerable amounts of debt, which in 1986 equated to $157 million, twice the shareholders' equity in the company.[74] AMC also began an overseas expansion including sites in Japan and Western Europe, including Britain.

United Artists Theatre Circuit's (UATC) rapid expansion in the 1980s was largely a result of acquisition rather than building, whilst its sites had an average of only 4.1 screens.[75] UATC had been formed in Maryland in 1926 and by 1988 it was the largest cinema circuit in the USA, with 620 cinemas with a total of 2702 screens,[76] the culmination of three decades of building and the acquisition of a number of regional circuits such as Gulf Theatres in 1986 and Georgia Theatres in 1987. UATC was a subsidiary of United Artists Communications Inc., which in 1986 had been taken over by Tele-communications Inc., only to be sold in 1992 to a management buyout, funded by investment bank Merrill Lynch Capital Partners. UATC continued to be the largest exhibitor in the USA through into the 1990s, and in 1995 had 2300 screens in 30 states, as well as sites in Asia, Mexico, and South America.[77]

Carmike's origins go back to the formation of a Georgia-based circuit called Martin Theatres in 1912, which Carmike's founder, C.L. Patrick joined after World War II and rose to become President. Patrick stayed with Martin Theatres after it was taken over by a conglomerate called Fuqua Industries in 1969, and, in 1982 having resigned from Fuqua, purchased Martin Theatres in 1982 when it was put up for sale. Patrick renamed the company Carmike, after his sons Carl and Michael, and began to expand the circuit rapidly. In the first six years, Carmike increased from 128 screens to 215, acquiring other circuits as well, especially in the southern states.[78] Its growth was based upon a strategy of building in smaller cities, especially where it could be the sole exhibitor, and keeping overheads down to the minimum. This meant that Carmike employed almost half of its staff on minimum wage, leased the majority of its buildings rather than own them, limited its range of concessions, and installed telephone and Internet booking at only a minority of sites.[79] It expanded throughout the 1990s via a strategy of continued acquisition but thereafter by building new cinemas, especially after 1998.

In just four years from 1973, National Amusements added more than 100 screens by a combination of expanding existing two- and three-screen cinemas and also building larger complexes of six to eight screens.[80] Interviewed in 1986, Redstone outlined the policy of the company when it came to location:

[w]e will only go with what we consider to be a triple-A location. That would mean on heavily travelled highways, with proximity to dense population. Since we generally will build only on land we own that puts a great deal of limitation on our expansion potential, which is probably good.[81]

By 1990, National Amusements was the eighth biggest circuit in the USA with 625 screens in 90 sites,[82] well behind giants like UATC and AMC; however, their cinemas were almost all multiplexes with an average of seven screens per site. Moreover, the company had been one of the first US companies to expand overseas and, unlike AMC for instance, it has retained its presence in those overseas markets. It entered Britain in 1988 (see Chap. 5) and South America in 1997, with multiplexes in Chile and Argentina. The other major circuits—Cineplex Odeon, General Cinema, and Carmike—all began to either expand their existing sites or build new, larger 8- to 12-screen complexes during the 1980s. In 1987, Cineplex Odeon, which at that point was less than eight years old, surpassed the Eaton Centre and built what was then the world's largest cinema complex, at the Universal Studios theme park in Los Angeles. With 18 screens, the $16.5 million complex could accommodate up to 6000 cinemagoers and would, according to Drabinsky, "bring the majesty back to moviegoing!"[83]

## "Movie Theater of the Future: The Home?"

In June 1977, JVC launched its Vidstar Video Home System, predicting that by 1980 as many as one million would be purchased.[84] In reality the total number of Video Cassette Recorders (VCRs) or Video Tape Recorders (VTRs) as they were sometimes called, sold by that year was 795,000, though this merely heralded an explosion in the number sold thereafter, which had reached 10.66 million per year in the USA just five years later in 1985, meaning 26.5 per cent of US homes owned one.[85] By 1990, that number had more than doubled, as 62.7 per cent of homes had a VCR and annual sales reached 12.9 million.[86] In a 1977 article provocatively entitled "Movie Theater of the Future: The Home?" Kopper argued that whilst the demand for films would continue in the near future, the major changes would be in the method of distribution, a "shift from seeing films in theaters to viewing them at home."[87] For many commentators then, the VCR was an existential threat to cinema exhibition, with 1985's 7 per cent decline in box-office and 12 per cent decline in admissions over 1984 confirming intuitively that threat.[88] For the exhibition industry, there were a range of questions but, according to Robert Selig, President of the National Association of Theater Owners (NATO), there was a key one: "what it's about is the capability of the movie theater to remain 'the place to go'."[89]

In 1986, GCC, at that time the largest exhibitor in the USA, completed a 21-month survey of 50,000 of its cinemagoers, which, amongst other things, concluded that VCR ownership had only a "3.5% negative impact on attendance."[90] GCC forecast that the net effect of video would be that the greater revenue profits accruing to the film suppliers (the studios) would benefit the industry as a whole.[91] Notwithstanding the perceived threat to film exhibition in the USA, as Balio pointed out, the 1980s were in fact "good to Hollywood" and cinemagoing, since the new home video technologies "stimulated demand for more motion pictures," enhancing "the status of the theatre in the distribution chain" as "the performance of a feature film at the box office established its value at the video store."[92]

The Hollywood studios had been reticent initially to make films available for video, which was a result of their desire to both control the distribution business and restrict copying of their films and television programmes. As part of their efforts to restrict copying, MCA Inc./Universal entered into a protracted legal battle with Sony Corporation that went all the way to the US Supreme Court in 1984. With the judges far from unanimous, and with an accompanying recommendation that Congress debate the issue, the decision eventually went in Sony's favour; the Supreme Court held that the "sale of the VTR's to the general public does not constitute contributory infringement of respondents' copyrights."[93] However, once the major studios engaged with the new market, they all came to rely heavily on the video sector to boost profits. For example, according to Gomery in 1980, the major Hollywood studios collected approximately $20 million from the worldwide sale of videocassettes; by 1986, this had risen to over $3 billion.[94] The studios' move into the video sector was important since they managed to establish as standard a distribution deal which gave them 75 per cent of the wholesale revenues from video—more than twice the revenue share they received off cinema and television distribution.[95]

In the 1980s, the small business nature of video rental in the USA, with grocery, or "mom and pop," stores stocking a range of titles for rent, began to change so that by the end of the decade this was increasingly the domain of large circuits and business concerns, with Blockbuster Video as the significant player. Blockbuster was started by David and Kevin Anderson in Dallas, Texas, in 1985 with one store, albeit one that stocked some 6500 titles and 8000 video cassettes.[96] The company grew phenomenally quickly, coming to occupy the corners of many streets in the USA and expanding overseas from the late 1980s. As if to epitomise the 'merger-mania' of the 1990s, Blockbuster and its 3300 stores worldwide were acquired by media giant Viacom in 1994 for $7.6 billion. Earlier in 1994, Viacom acquired Paramount Studios for $10 billion, making Paramount the first film studio to be controlled by a company whose primary business was the ancillary film market.[97] Significantly, the Chairman and largest stockholder of Viacom was Sumner Redstone, whose privately owned National Amusements operated the Showcase multiplex cinema circuit.

The entry of DVD into the market in 1997 was dramatic, with sales of the players increasing from 350,000[98] in its first year to 22.1 million in 2003, making DVD the fastest selling technology of all time.[99] When one includes Canada, the penetration of DVD in North American households was 57 million in 2003, with the average household renting 34 DVDs a year compared to just 7.5 in Europe.[100] The development of the VCR and DVD subsequently resulted in the attraction of film studios to the producers of the hardware, what were increasingly referred to as synergies. In November 1989, for instance, Sony purchased Columbia Pictures for $3.4 billion in a deal that *Variety* characterised as fitting into "Sony Corp.'s future plans as easily as a videocassette slides into a VCR."[101] When Sony took control of Columbia Pictures from The Coca-Cola Company it also acquired Tri-Star Pictures and the Loews Theater circuit (Loews Theatre Management Corp.), which had been purchased by Tri-Star in 1986. Sony was thus a fully vertically integrated company with a circuit of cinemas numbering 175 and containing over 1000 screens, when in 1994 it changed the marquee names to Sony Theatres. The Sony Lincoln Square, located on Fifth Avenue in New York, opened in 1995, was a 12-screen, 3600-seat multiplex, taking up the first seven floors of a 50-floor office block and containing an IMAX screen, which was at that point the largest in the USA.[102] Albanese wondered if it was "the shape of things to come?"[103] and *The Film Journal* considered it as providing "a view of the future."[104] Sony's entry into the film business, echoed by Japanese electronics company Matsushita's purchase of MCA Inc. (owner of Universal Studios) in 1990 for more than $6 billion, was the culmination of a decade of dramatic change in the culture and economics of the entertainment industry in the USA.

## VERTICAL INTEGRATION AND SYNERGY—"THERE'S NO BUSINESS LIKE THE EXHIBITION BUSINESS"

The growth of the major studios in the 1980s and early 1990s was in part the result of mergers and conglomeration, and a renewed re-examination of the benefits of vertical and horizontal integration. Significantly, during the 1980s many of the major studios and their parent companies acquired cinema circuits, as the *Paramount Decrees* came under greater pressure. As Holt pointed out, "[b]y the end of Reagan's tenure in office, the major studios or their parent companies owned almost the same percentage of the country's theaters that the 'Big Five' held prior to the *Paramount* consent decrees."[105] The attitude seemed to be that the continued economic health of the studios rested on the maximisation of profits from a wide spectrum of mass-media enterprises.[106] The buzzword on the lips of studio executives across the USA was "synergy"—"[a]s in a good marriage, each partner would bring qualities that when combined would magically create something better than either could achieve alone."[107] Moreover, as Fabrikant observed, "[i]n the giddy 80's, every studio executive, an investment banker by his side, was primed to bid for theater circuits. 'Synergies', they said—combine studios with theaters, and inevitably there would be synergies."[108] The notion of synergy, which for film studios meant a process of vertical integration, was also the basis for the practice of horizontal integration. Here, the traditionally high-risk activities of filmmaking were complemented with a broader range of leisure businesses, so that the recouping of film production costs was spread out across a series of related activities. These could include soundtracks, computer games, television, video, toys, merchandising, and tourism via theme parks. Cineplex Odeon was a good example. At the same time as it was expanding the exhibition arm in the first half of the 1980s, Cineplex Odeon was seeking to expand the company into other film-related areas such as film production, post-production, and distribution. The distribution arm of the company, initially established in order to provide their own circuit with foreign-language and arthouse films, became one of the largest in Canada obtaining rights to distribute films not only in cinemas but also in the burgeoning video market and on cable TV.[109] The efforts to become a vertically integrated company as well as its ambitious building programme meant that Cineplex Odeon needed additional finance, which they obtained from the entertainment conglomerate MCA Inc., owner of Universal Studios in 1986, in the form of $200 million, in exchange for 49.7 per cent of Cineplex Odeon's stock.

Though much horizontal integration was concerned with US domestic re-organisation and mergers, film producers sought to strengthen their profit base, lessen their financial exposure, and reduce their debt load by globalising their operations. For many of the major studios the ability to dominate markets at home and abroad had been based on a resurgence of interest in vertical integration, made possible by the relaxing of enforcement of anti-trust laws during the 1980s, and was part of a general trend to deregulation or liberalisation on the part of the Reagan administration. In 1980, *Variety* quoted Senator Robert Packwood who said that the "No. 1 concern of the Senate Commerce Committee next year will be deregulation of the telecommunications industry."[110] It should not, however, simply be assumed that vertical integration and the purchase of exhibition circuits, by the studios—themselves the major distributors—did not come under scrutiny by the regulators. In July 1986, the antitrust division of the US Justice Department made it clear that it was "monitoring" the recent trend for acquiring exhibitors, though it acknowledged that none of the deals had thus far contravened the appropriate sections of the *Paramount Decrees.*[111] As Pryor observed Hollywood's new theme song was: "There's no business like the exhibition business."[112]

A significant test of the antitrust laws had come in 1985, when Columbia Pictures had acquired the Walter Reade circuit and in 1986, when MCA Inc./Universal purchased 49.7 per cent of Cineplex Odeon (see above). In reality, the studios had been vertically integrated for a number of years since they both produced and distributed films. The entry into exhibition constituted "merely a 'downward' extension of vertical integration."[113] As we have seen, the *Paramount Decrees* actually barred only three studios from acquiring cinemas—Warner Bros., 20th Century Fox, and MGM. The National Association of Theater Owners (NATO) was initially opposed to membership by these new circuit owners; however, in late 1986 in the light of the Justice Department's view that the Paramount and MCA Inc./Universal deals did not violate the *Paramount Decrees*, it reversed its opposition and allowed these companies to be members.[114]

It became clear to many, especially amongst film industry executives, that the Reagan administration's market-oriented ideology was more than simply about deregulation. As Holt argued, "[i]t is important to understand deregulation in the Reagan era not as the mere absence of regulation, but instead as the presence of a politicized and carefully crafted government stance of support for marketplace policies. Definitions of regulation and corporate culture were undergoing tremendous change during this time."[115] This laissez-faire attitude to regulation was continued by Republican President George Bush, culminating in the deregulation of the media sector under Democratic President Clinton and the Telecommunications Act of 1996. According to Schatz this:

> opened the way for cross-media ownership of film, television, cable, music, publishing, and other media and entertainment interests—thus propelling the rise of a cadre of media giants that would integrate several once distinct media industries (movies, television, cable, music, publishing, et al.) into a worldwide entertainment industry with the film studios at the epicenter, and with "filmed entertainment" as its key commodity.[116]

The major studios were beneficiaries of decades of increasing domination of, initially, Hollywood cinema and, latterly, popular television. Along with the changes in the organisation of global capitalism and political systems, the commercialisation of broadcasting systems worldwide has led to the globalisation of the film and television industry.

As we have seen, the key observation here was that of the 'secure home base' which for Hollywood and cinema in the USA has been the bedrock on which its domination has been built. Initially, Hollywood's domination overseas was due to its large home market that assured a film's profitability before it was sold abroad, usually at a rate that was more than competitive with domestic film production in other countries. Although the numbers of cinemas dwindled in most overseas countries (or 'territories'), the demand for films in other mediums, notably television from the 1960s and video from the 1980s, increased. Moreover, in the period since the mid-1980s, with the development of multiplexes and other cinema upgrading, there has been a continued growth in the overseas markets for Hollywood. Despite this growth in cinemas on the eve of 'The 1996 Act' the proportion of revenues for American film companies derived from foreign theatrical rentals, which used to be almost the total, now accounted for approximately a quarter, with video becoming the largest revenue stream.[117] By the end of the 1980s, according to a report by Communications Equity Associates Inc., the studios owned some 17 per cent of US cinemas,[118] and were, argued Britain's Cinema Exhibitors Association, testing "the advantages of vertical integration through exhibition ownership."[119]

### THE MULTIPLEX IS ESTABLISHED—THE MOVIE PALACE GOES THE WAY OF THE HORSE AND BUGGY

Interviewed in *BoxOffice*, A. Alan Friedberg, president of Sack Theatres and chairman of the National Association of Theater Owners (NATO), commented that the "old movie palace will come back when the horse and buggy does. The multiplex is the most economical way for the exhibitor to operate, and it offers the most diversity of choice for the patron."[120] At the start of the 1990s, there were, broadly speaking, two prevailing views about the state of exhibition in the USA. The first was that the market was saturated or at best "mature"; whilst the second was that there was significant potential for growth not just in terms of more screens but also in an enhancing of the cinemagoing "experience." As Acland observed, it seemed that "no consensus existed whether North America was over- or underscreened; however, there seemed to be an agreement that the continent was in some way inadequately screened, that its screens were in the wrong location and were equipped with archaic technology and design."[121] Speaking in 1996, Howard Lichtman from Cineplex Odeon identified "three key factors" driving expansion:

> [i]nvestment by the studios in producing more product, a direct result of the growth of ancillary and international markets; a depressed real estate market, which makes building and operating more viable; and better deals from [shopping] malls, which have begun to see the value of cinemas as a traffic-builder.[122]

Lawrence Cohn set out the impact of the multiplex after the decade of expansion in 1990. They were: firstly, that converted and twin cinemas were cramped and that the new purpose-built multiplexes were more comfortable and "state of the art"; secondly, multiplexes have led to huge increases in the P&A costs for distributors; and thirdly, that multiplexes have resulted in producers increasingly focussed on mass market films rather than speciality films.[123] This meant that there existed a symbiotic relationship between the major exhibitors and the major studios, themselves increasingly part of ever larger transnational media concerns. One of the outcomes of this symbiotic relationship was a sense that the audience was there and the films were there, therefore it was up to exhibitors to provide ever more screens to accommodate them. As the 1980s gave way to the 1990s, increasingly the multiplex began to expand from 8–10 to 12–16 screens and more, with the major circuits growing and accelerating openings.[124] As Larry Gleason from Paramount Pictures observed though, success was "going to depend heavily on the natural selection—by cinema audiences—of one multiplex over its neighbour."[125]

## NOTES

1. William Paul, "The K-mart Audience at the Mall Movies," *Film History* 6, no. 4 (Winter, 1994): 487–501, 493.
2. Ibid., 491.
3. Jack Valenti, "Theatre Owners and Changing Audience," *Screen International* 285 (28 March, 1981), 6.
4. "AMC in Huge Expansion Plan for 1971 With 70 Auditoriums in 13 Cities," *BoxOffice* 98, no. 26 (12 April, 1971), 4.
5. John W. Quinn, "American Multi at 457; Formula: Follow Malls," *Variety* 285, no. 8 (29 December, 1976), 20.
6. Douglas Gomery, *Shared Pleasures: A History of Movie Presentation in the United States* (London: British Film Institute, 1992).
7. Richard Albarino, "Just How Many Film Sites Are There?," *Variety* 277, no. 9 (8 January, 1975), 7.
8. Quoted in Candace Burke-Block, "M.B.C. Construction Head Cadiff States 'Twinning Is Good, Multiplexing Is Better'," *The Film Journal* 84, no. 4 (9 March, 1981), 35.
9. "Movie Theater Gets Cut to Size," *Business Week* (14 March, 1970), 29.
10. Charles. F. Rouse III, "Will Faltering Economy Curtail Theater Growth?," *BoxOffice* 116, no. 18 (5 May, 1980), 11.
11. Quoted in Barbara Stones, *America Goes to the Movies: 100 Years of Motion Picture Exhibition* (Hollywood: National Association of Theater Owners, 1993), 222.
12. "AMC Opens 7-Theatre Complex in Toledo," *BoxOffice* 10, no. 19 (21 August, 1972), ME2.
13. "AMC Bows Nine New Screens in Toledo," *Variety* 320, no. 12 (16 October, 1985), 8.

14. "Cobb Praises Christie Equipment in Use at Chain's Cinema City 8," *The Independent Film Journal* 81, no. 13 (1 October, 1978), 33.
15. "Cobb's 'Octoplex': Eight Theatres, Two Floors of Film, Food and Fun," *BoxOffice* 113, no. 24 (18 September, 1978), A17.
16. Philip Turner, *Cineplex Odeon: An Outline History* (St. Paul's Cray: Brantwood Books, 1998).
17. David Linck, "Canada's Cineplex: A New Idea Goes South," *BoxOffice* 117, no. 11 (1 November, 1981), 74.
18. Turner, *Cineplex Odeon*, 5.
19. J. W. Agnew, "Cinema + Complex = Cineplex," *BoxOffice* 115, no. 3 (23 April, 1979), A10.
20. "With Cineplex Plus Odeon, Garth Drabinsky Adds Up to Major Theater Force," *Variety* 317, no. 4 (21 November, 1984), 62.
21. "2 Boston-Based Exhibs Offer Upstate N.Y. Sites; Are to Add 24 Screens," *Variety* 314, no. 10 (4 April, 1984), 30.
22. Debra Berliner, "Cinema Centers Corp. Projects 150 New Screens in Five Years," *The Film Journal* 87, no. 7 (1 July, 1984), 31, 43.
23. Quoted in "N.Y. Now Cinema Centers' Top Market; Adding Albany Ten-Plex," *Variety* 310, no. 20 (20 April, 1983), 10.
24. "Movie Attendance Hits 11-year Peak," *BoxOffice* 115, no. 32 (5 November 1979), 1.
25. Dodona, *Moviegoing* (Leicester: Dodona Research, 1999), 20.
26. "Exhibs Bid 'Adieu' To Mini-Cinemas," *The Film Journal* 85, no. 18 (23 August, 1982), 7.
27. Paul, "The K-mart Audience at the Mall Movies."
28. MPAA, *US Economic Review. Motion Picture Association of America*, 1998, 9.
29. Dodona, *Moviegoing*, 31.
30. Dodona, *Moviegoing 5* (Leicester: Dodona Research, 2005), 20.
31. Mark Jancovich and Lucy Faire, with Sarah Stubbings, *The Place of the Audience: Cultural Geographies of Film Consumption* (London: British Film Institute, 2003), 190.
32. Geoff King, *New Hollywood Cinema: An Introduction* (London: I.B. Tauris, 2002).
33. "In Focus: The Faith of the Decrees," *The Film Journal* 92, no. 6 (1 July, 1989), 3.
34. Mark Batey, "Blockbusters and Beyond," *Cinemabusiness* 30 (December, 2005), 30.
35. See Marco Cucco, "The Promise Is Great: The Blockbuster and the Hollywood Economy," *Media, Culture & Society* 312 (2009): 215–30.
36. "Worldwide Cinema: Weakness Hidden by Record Growth," *Screen Digest* (September, 1998), 208.

37. Dodona, *Moviegoing*, 20
38. Dodona, *Cinema Industry Research: USA & Canada* (Leicester: Dodona Research, 2018), 1.
39. "Worldwide Cinema: Poor Product Fails the Multiplex Boom," *Screen Digest* (September, 2000), 283.
40. Dodona, *Moviegoing*, 12 and Dodona, *Cinema Industry Research: USA & Canada*, 1.
41. Jeremy Floyd Spracklen, *The General Cinema Northpark I & II: A Case Study of a Third Generation Movie Theater* (Unpublished MA History thesis presented to The University of Texas at Arlington, 2015).
42. Robert Venturi, Denise Scott Brown, and Steven Izenour, *Learning from Las Vegas: The Forgotten Symbolism of Architectural Form* (Cambridge: MIT Press, 1977).
43. Quoted in Clyde R. McKinney, "Practical Theatre Design: Back to the Future," *BoxOffice* 123, no. 3 (1 March, 1987), 74.
44. Paul, "The K-mart Audience at the Mall Movies," 493.
45. Garry Edgerton, "The Multiplex: The Modern American Motion Picture Theater as Message," *Journal of Popular Film and Television* 9, no. 4 (Winter, 1982): 158–65, 163.
46. Kevin Lally, "New Designs: Exhibitors Strive for Old-Time Luxury," *The Film Journal* 89, no. 6 (1 June, 1986), 23.
47. Paul, "The K-mart Audience at the Mall Movies," 494.
48. George Ritzer, *The McDonaldization of Society*, Revised Edition (Thousand Oaks: Pine Forge Press, 1996), 1.
49. See Stuart Hanson, "Spoilt for Choice? Multiplexes in the 90s," in *British Cinema of the 90s*, ed. Robert Murphy (London: British Film Institute, 2000).
50. Tom Matthews, "Multiplex Movie Palaces?" *BoxOffice* 124, no. 3 (1 March, 1989), 24.
51. Quoted in Richard Natale, "Designed to Entertain," *Variety* 373, no. 7 (4 January, 1999), 83.
52. Ibid., 83.
53. Edwin Heathcote, *Cinema Builders* (Chichester: Wiley-Academy, 2001), 151.
54. Ibid., 151.
55. Brian Holberton, "Today's Lobby," *Film Journal International* 103, no. 11 (1 November, 2000), 72.
56. Joseph Kupfer, "From Edifice to Device: Architecture of Disengagement," *Technology in Society* 12 (1990): 319–32, 325–6.
57. Quoted in Braden Phillips, "Regal at 10: Stadium Seating Tops List of Theater Design Priorities," *Variety* 374, no. 3 (8 March, 1999), 36.

58. Quoted in Jim Robins, "Exhibs See a Trend to Larger Houses: Big Screen to Fight Homevideo," *Variety* 308, no. 1 (4 August, 1982), 3, 28.
59. On 1 December 1988, *BoxOffice* (124, no. 12: 18–27) presented its first annual survey of the "Giants of Exhibition" with a profile of the top 10 largest American exhibitors.
60. "Special Report: Strategic Trends in Theatrical Exhibition: The Independent Exhibitor's Survival Guide," *BoxOffice* 124, no. 7 (1 July, 1989), 14.
61. Will Tusher, "General Cinema Plans to Add 300," *Variety* 317, no. 6 (5 December, 1984), 7.
62. Kevin Lally, "General Cinema Unveils a New Look," *The Film Journal* 88, no. 1 (1 January, 1985), 79.
63. Ralph Kaminsky, "14-Theatre Beverly Center Cutting Tape This July on 'Cineplex Day'," *The Film Journal* 85, no. 15 (28 June, 1982), 4.
64. Turner, *Cineplex Odeon*.
65. Kaminsky, "14-Theatre Beverly Center Cutting Tape."
66. John Harkness, "Cineplex Faces More Financial Problems," *Screen International* 390 (16 April, 1983), 12.
67. "U.S. and Canada Theatre Circuits Ranked by Size," *Variety* 338, no. 4 (31 January, 1990), 20.
68. Turner, *Cineplex Odeon*, 18.
69. Tom Matthews, "Theatre Profile: AMC's Century 24," *BoxOffice* 124, no. 2 (1 February, 1988), 14.
70. "AMC Plans Further Expansion in Century City Shopping Centre" *Screen International* 544 (19 April, 1986), 6.
71. "Varied Array of Firstruns and Art Pictures Fills Century City 14," *Variety* 328, no. 12 (14 October, 1986), 30.
72. Quoted in Matthews, "Theatre Profile: AMC's Century 24," 12.
73. "U.S. & Canada Theater Circuits Ranked By Size," 20.
74. A. B. Block, "What Makes Stanley Borrow? Stan Durwood's AMC Entertainment Is Loaded with Debt and Costly Leases. So Why Does the Stock Sell for 27 Times Earnings?" *Forbes*, 22 September, 1986, 54.
75. "U.S. & Canada Theater Circuits Ranked By Size," 20.
76. "Ten Giants: United Artists," *BoxOffice* 124, no. 12 (1 December, 1988), 18.
77. Leonard Klady, "United Artists Theatre Circuit Inc.," *Variety* 360, no. 4 (28 August, 1995), A58.
78. "Ten Giants: Carmike," *BoxOffice* 124, no. 12 (1 December, 1988), 23.
79. Dodona, *Moviegoing 4* (Leicester: Dodona Research, 2004).
80. "A Redstone Milestone," *Film Journal International* 99, no. 1 (1 November, 1996), 178.

81. In Jim Robbins, "Redstone Bares Strategy in Building Circuit Empire," *Variety* 323, no. 11 (9 July, 1986), 60.
82. "U.S. & Canada Theater Circuits Ranked By Size," 20.
83. Jimmy Summers, "Cineplex Odeon Launches a Dazzling New Flagship," *BoxOffice* 123, no. 9 (1 September, 1987), 8.
84. *Backstage*, "JVC Intros Home Video System," 18, no. 24 (17 June, 1977), 29.
85. "World VCR Markets: Back into Growth," *Screen Digest* (June, 1993), 132–3.
86. Ibid., 133.
87. Phillip Kopper, "Movie Theater of the Future: The Home?," *American Film* 3, no. 3 (1 December, 1977), 17.
88. "MPAA Reveals US and Canadian Box Office Grosses Down 7% from 1985," *Screen International* 537 (1 March, 1986), 16.
89. Robert E. Selig, "Exhibs Think Big as Filmgoers Have Small Home Screens," *Variety* 321, no. 11 (8 January, 1986), 34.
90. Frank Segers, "Say VCR Effect on TIX Sales Peaking," *Variety* 321, no. 12 (15 January, 1986), 36.
91. Ibid.
92. Tino Balio, "Adjusting to the New Global Economy: Hollywood in the 1990s," in *Film Policy: International, National and Regional Perspectives*, ed. Albert Moran (London: Routledge, 1996), 23.
93. U.S. Supreme Court. *Sony Corp. v. Universal City Studios, Inc.*, 464 U.S. 417 (1984).
94. Douglas Gomery, "Hollywood's Hold on the New Television Technologies," *Screen* 29, no. 2 (Spring, 1988): 82–9, 88.
95. For an account of the ways in which the Hollywood studios derive revenue from a variety of distribution channels, see Janet Wasko, *How Hollywood Works* (London: Sage, 2003).
96. See Paul McDonald, *Video and DVD Industries* (London: British Film Institute, 2007).
97. Colin Brown, "Viacom Constructs a 'Global Powerhouse'," *Screen International* 945 (18 February, 1994), 1.
98. Marc Graser, "Jolly Times for DVDs," *Variety* 373, no. 5 (14 December, 1998), 20.
99. Paul Taylor, "US Consumer Electronics Sales to Top $10bn," *Financial Times*, 8 January, 2004.
100. "Worldwide DVD Markets in Position," *Screen Digest* (November, 2004), 336.
101. Charles Kipps, "Sony and Columbia: A Tidy Case of Hardware Meeting Software," *Variety* 336, no. 11 (27 September, 1989), 5.

102. In January 2018, despite being highly profitable the cinema closed as the landlord did not renew the lease. See Thomas Beller, "The Death of a Movie Theatre," *The New Yorker*, 6 September, 2018.

103. Alex Albanese, "The Shape of Things to Come?" *BoxOffice* 131, no. 4 (1 April, 1995), SW16.

104. "Sony Lincoln Square Theatre Provides View of the Future," *The Film Journal* 98, no. 22 (1 March, 1995), 34.

105. Jennifer Holt, "In Deregulation We Trust: The Synergy of Politics and Industry in Reagan-Era Hollywood," *Film Quarterly* 55, no. 2 (Winter, 2001): 22–9, 22.

106. See Gomery, *Shared Pleasures* and Douglas Gomery, "Hollywood corporate Business Practice and Contemporary Film History," in *Contemporary Hollywood Cinema*, eds. Steve Neale and Murray Smith (London: Routledge, 1998).

107. Calvin Simms, "Synergy': The Unspoken Word," *New York Times*, 5 October, 1993.

108. Geraldine Fabrikant, "All About/Movie Theaters; The Crunch Comes to Cinemas," *New York Times*, 25 November, 1990.

109. Joseph Lampel and Jamal Shamsie, "Darth's Dominion: A Case Study of the Cineplex Odeon Empire," *Cinema Canada* (October, 1989): 17–21.

110. "Media Deregulation Is Top Concern of Senate Committee," *Variety* 301, no. 4 (26 November 1980), 1.

111. Tusher Will, "Justice Dep't. Has Eye on Distributors: Collusion's What it's Watching for," *Variety* 324, no. 2 (6 August, 1986), 5.

112. Thomas. M. Pryor, "The New Exhibition Mania," *Variety* 32, no. 2 (6 August, 1986), 5, 31.

113. Ibid., 29.

114. "NATO Okays Distrib Exhibs," *The Film Journal* 89, no. 9 (1 September, 1986), 44.

115. Holt, "In Deregulation We Trust," 26.

116. Thomas Schatz, "The Studio System and Conglomerate Hollywood," in *The Contemporary Hollywood Film Industry*, eds. Paul McDonald and Janet Wasko (Oxford: Blackwell, 2008), 27.

117. Balio, "Adjusting to the New Global Economy."

118. "Strategic Trends in Theatrical Exhibition: The Independent Exhibitor's Survival Guide," *BoxOffice* 124, no. 7 (1 July, 1989), 15.

119. Ibid., 15.

120. A. Alan Friedberg, President of Sack Theatres and chairman of the National Association of Theater Owners (NATO), quoted in Alexander Auerbach, "Despite the Christmas-Season Chill Circuit Heads Remain Optimistic," *BoxOffice* 118, no. 4 (1 April, 1982), 22–3.

121. Charles Acland, "Cinemagoing and the Rise of the Megaplex," *Television and New Media* 1, no. 4 (2000): 375–402, 387.
122. Ana Maria Bahiana, "All the World's a Screen," *Screen International* 1047 (1 March, 1996), 20.
123. Lawrence Cohn, "Fewer Plexes but More Multi," *Variety* 341, no. 3 (29 October, 1999), 1.
124. See Charles Acland, *Screen Traffic: Movies, Multiplexes and Global Culture* (Durham and London: Duke University Press, 2003).
125. Quoted in John Hazelton, "Drowning in a Sea of Screens," *Screen International* 734 (9 December, 1989), 17.

# CHAPTER 4

# From Multiplex to Megaplex

As the 1990s began, the multiplex had become the de facto norm for cinemas in the USA, as throughout the 1980s the number of screens increased rapidly—55 per cent between 1980 and 1988, which equated to 7700 screens mostly in new multiplexes—though the increase in cinemagoing was only 6 per cent over the same period.[1] The increase in multiplexes also had a profound impact upon distribution patterns, whilst building would proceed apace right through to the new millennium, as three major circuits—American Multi-Cinema (AMC), Regal Entertainment, and Cinemark—emerged during the 1990s as the dominant forces in exhibition. The change in the complexion of the exhibition industry and the disappearance of some of the dominant circuits from the 1980s would be the result of a series of bankruptcies and mergers, in part precipitated by indebtedness as a result of the enormous costs involved in the building of ever larger multiplexes, including giant 20- to 30-screen 'megaplexes,' and the updating of older sites. Throughout the 1990s and into the new millennium, admissions grew steadily so that by 2002 the USA recorded 1.625 billion, which was the highest since 1957.[2] Having peaked in 2002, admissions then declined until 2017, when the figure of 1.126 billion was the lowest this century.[3] In the face of competition from home video in the 1990s, the industry had to contemplate that of streaming services like Netflix in the 2000s, with the concomitant challenges to the release window system, that operated in agreement with the studios, which hitherto controlled distribution.

© The Author(s) 2019
S. Hanson, *Screening the World*,
https://doi.org/10.1007/978-3-030-18995-2_4

## THE MEGAPLEX—'WELCOME TO THE FUTURE OF MOVIEGOING!'[7]

By the early 1990s the major circuits began a programme of replacing many of their smaller sites with larger multiplexes, a significant example of which was AMC's Ward Parkway 12 in 1992, built on the site of the original Parkway I and II Twin and subsequently expanded to 22-screens in 1995.[4] The expansion of the Ward Parkway 12 was indicative of the key development in multiplex building in the mid-1990s: that of the megaplex.[5] These were multiscreen complexes characterised as having 16 screens or more, auditoria with stadium seating—high backed seats installed on steep risers—and the latest digital sound.[6] In canvassing views of the new megaplexes from amongst industry insiders, *Film Journal International* highlighted two categories of megaplex: complexes of 16–18 screens, with 3500 seats and 40,000 square feet, and complexes of 20 screens or more, with 5000+ seats and 75,000 square feet or more.[7]

AMC had heralded these developments with the announcement in 1994 that it had begun construction on an 85,000-square-foot, 24-screen complex, costing $12–13 million, at the Stemmons Crossroads development in Dallas, Texas. The Grand 24, as it was called, was the largest cinema in the USA when it opened in May 1995, and had computerised ticketing, stadium seating, Sony Dynamic Digital Surround in all auditoria, a 12-window ticket booth, and a 7000-square-foot main lobby.[8] AMC trumpeted it as the "future of moviegoing!"[9] In addition to the main concession stand, each of the two 12-screen wings in the complex contained a secondary lobby area and further concessions stands. There were 4600 seats in auditoria which ranged from 120 to 420 seats whilst the adjacent car park was 13 acres in size.[10] According to Durwood, the complex represented "not only the future of AMC but the comprehensive entertainment centre of tomorrow."[11]

Stadium seating was seen by many exhibitors as a significant attraction to audiences, not least because of the improved sightlines offered over traditional sloped auditoria. However, for the circuits, the costs of accommodating stadium seating were considerably higher, since with wider seats and wider row spacing, each row needed to be 12–14 inches higher than one below it, which required taller buildings.[12] As the seats at the back of the auditoria got closer to the ceiling, this also necessitated alterations to the sound system so as to disperse the sound and not deafen patrons on the back rows. Moreover, though stadium seating satisfied existing building codes and the Americans with Disabilities Act (ADA) they were less popular with the elderly and disabled, with many circuits fearing that costs might be increased by making future designs more accessible or worse still, necessitating costly alteration to existing cinemas.[13]

The logic of the megaplex was commercial, since they were seen as generating more revenue than smaller multiplexes though the costs of opening were considerably higher. AMC, like other circuits, began to off-set the costs of opening these giant complexes by selling sites to real estate investors and leasing them back. Once opened, though, the megaplex offered increased efficiencies and substantial economies of scale over multiplexes, with claims of greater average attendances, higher average revenue per customers, and greater margins as measured by operating cash flow before rent.[14] Regal Cinema's head Michael Campbell saw megaplexes as "wonderful assets and models of efficiency" but they could also "siphon moviegoers from existing theatres like a huge vacuum cleaner."[15] Moreover, Campbell also observed that the cinemas being "cannibalized" were not "old, tired or small theatres, but theatres of eight, ten or even twelve screens that might be only a few years old" and which when they were built "were expected to be sound 15 to 20 year investments."[16] Nonetheless, as Joost Bert, head of Decatron International (see Chap. 8), announced, "[m]egaplex, ultraplex or giant-plex—it means more seats and more opportunities for people to leave their homes and go out and experience a joyous and entreating time."[17]

At Stemmons Crossroads, AMC's Grand 24 megaplex was, according to AMC's Philip Singleton, intended to be an "example of operating leverage. When they hit the margin, beyond the breaking point, they hit in a very big way."[18] Within three years of opening The Grand 24, AMC had 38 megaplexes containing 800 of the circuit's 2312 screens[19] with the 30-screen, 5700-seat Ontario Mills shopping complex in Ontario Mills, California, opened in December 1996, the largest cinema in the USA at that time. The inclusion of extra wide, reclining seats in the company's new sites and the conversion of some existing auditoria had been particularly successful. By 2013, 35 of AMC's cinemas had been converted and although the wider seats meant a 64 per cent reduction in capacity, attendance at these complexes was up 80 per cent.[20] Between 2011 and 2017, AMC's cinemas that switched to recliners averaged 57 per cent growth in attendance and 225 per cent increase in revenues.[21] AMC's use of online booking and reserved seating meant that it was able to move films between auditoria as a result of real-time reservation information.[22]

AMC was not the only company opening the new megaplexes, since as Travis Reid from Sony Theatres observed, "I think the whole industry went to school on the AMC Grand."[23] In November 1995, Southern California–based circuit Edwards opened a 21-screen megaplex—dubbed "The Big One"—at the Irvine Spectrum Center in California.[24] The 158,000-square-foot complex, the biggest of its kind in the world at that time, cost $27 million, had only the second 3D IMAX screen in the country, and could seat 6400 people. According to its architect Marios Savopoulos, the new megaplex reflected:

> the new trend in the industry, which is to go into the malls and commercial areas and bring pedestrian activity into the area and create urban plazas. A lot of these are exterior spaces, and I think that's going to be the trend in the next ten years.[25]

Edwards' subsequent megaplex, opened in Ontario, California, in March 1997, was a 22-screen, freestanding design similar to that in Irvine; however, its position was significant: 300 yards from AMC's 30-screen Ontario Mills complex. The head-to-head competition, at that time the greatest concentration of screens in one location, was predicated on the surrounding population of 1.6 million people being able to sustain two giant cinemas. As La Franco observed both "Durwood and Edwards are going for luxury and glitz. Softer seats. More legroom. Steeply sloping, stadium-style seating. Larger screens. Fresh-made pizza. Flavored coffees. Candy by the pound. And above all, powerful digital sound systems."[26] Nonetheless, the box-office for both did not seem unduly affected by the proximity of the other and, whilst there were some differences in comfort (AMC's legroom was better) and film presentation (Edwards' megaplex offered Dynamic Theater Sound digital sound in all auditoria whilst AMC equipped only 14 of its screens with Sony Dynamic Digital Surround sound), the decision as to which complex to see a film was negated by the fact that the exhibitors largely divided the available films between them.[27] As of 2018, both complexes were still open for business.

Cinemark, which opened a 16-screen complex within the Universal Mall in a suburb of Detroit in 1991, opened its first freestanding 17-screen megaplex in Dallas in the same year as AMC opened the Grand 24 in the same city. It rivalled AMC in terms of its aggressive growth, especially with its branding of its megaplexes as 'Tinseltown USA.' Cinemark's rise to become one of the 'Big Five' was relatively quick, with the company coming into existence only in 1984 when Lee Roy Mitchell and Paul Broadhead combined to purchase a small circuit of cinemas from Consolidated Theatres Co. in Utah. Though Broadhead was a property developer with interests in shopping malls, Mitchell, like Stanley Durwood and Richard Smith, had been in the exhibition business since the late 1950s when he took over his father's small, Texas-based circuit and began growing the business by acquiring other sites in Texas. Throughout the 1960s and 1970s, Mitchell adopted a pattern of purchasing sites, building up the business and subsequently selling them on so that by the time he purchased Consolidated in 1984 and formed Cinemark with Broadhead, Mitchell had sold the last of his circuit of 50 cinemas.

For film distributors and the exhibition circuits, the megaplex offered the flexibility to satisfy demand for the growing number of blockbuster films being produced by the major studios from the mid-1990s onwards; what one cinema owner called "a programming dream come true."[28] Megaplexes could, it was claimed, provide a wide range of start times to accommodate whoever wanted to see a particular film, as well as being able to retain films for longer. Moreover, according to Dodona, the extra screens allowed for subruns of films which had finished their first run but which could be showed at times when there were fewer big new releases.[29] A benefit here was that since the terms of a rental agreement gave the distributors a larger percentage of admissions at the start of film's run but more to the exhibitor as the films play on, then having more screens meant films could be held over. Like the stated appeal of the multiplex, the megaplex offered even greater 'choice' of film and screening times. This notion of choice and size echoed other retail developments in the USA. As Melnick and Fuchs pointed out, "in a land where, today, you can buy a 24-pound jar of mayonnaise inside a 150,000-square foot warehouse, a theatre housing 24 auditoria seems congruous with a movement to grow retails spaces ever bigger."[30]

The arrival of the megaplex in 1995 heralded an increase in the number of first-run screens by some 6 per cent, which boosted box-office by 8 per cent; however, even though in 1996 13 films grossed $100 million there was no significant increase in the numbers of people going to the cinema.[31] What was happening, it seemed to many, was that these new megaplexes were merely redistributing attendances. National Amusements' Sumner Redstone argued that unless they were built in a market that was under-served by existing cinemas they may "only cannibalize existing business."[32] Unlike many rivals, which began to construct 16–24 screen megaplexes, National Amusements generally favoured 10–14 screen sites but with larger auditoria and more seats; indeed, Redstone argued that he could see no instance to support the building of anything greater than a 16-screen cinema.[33] This was predicated on the availability of film product, since, according to Vice-president Shari Redstone, many megaplexes were simply showing the same film on a number of smaller screens and "they are not a cookie-cutter answer to motion picture exhibition," adding that "if we felt that was the way to go, we would have pioneered it five years ago."[34] Nevertheless, even National Amusements were tempted subsequently to build larger complexes in some locations, such as those in Louisville, Kentucky; Ypsilanti, Michigan; and Revere, Massachusetts (all with 20 screens).

For many industry observers, the less-trumpeted intention of the megaplex was to strengthen the position of exhibitors in terms of rental negotiations with distributors. Nevertheless, what seemed to have happened was that these giant cinemas were, according to Dodona, about "capturing market share" and by "devoting a large number of screens to blockbuster releases" megaplexes could "suck up demand much faster than old line multiplexes."[35] However, cautioned Cineplex Odeon's Howard Lichtman, "the flip side is that when there is a dearth of product, there are a lot of mouths to feed."[36] Notwithstanding the risks of further over-screening and saturation, the six biggest cinema circuits—AMC, United Artists, Carmike Cinema, Cinemark, Loews Cineplex (formally Cineplex Odeon), and Regal Entertainment—had, according to Smith, spent $2.8 million in just three years from 1996, most of it borrowed.[37] It is perhaps no surprise that the boom in megaplex building corresponded with a boom in the financial markets in the 1990s, with low interest rates and plenty of capital available for investment, which encouraged such a great deal of borrowing on the part of the major circuits.

By 2000, five years after AMC's Grand 24 opened, there were 405 megaplexes (16 screens or more) and 1478 multiplexes (8–15 screens) in the USA out of a total of 7421 sites.[38] No other decade saw such an increase. If one considers the leading circuits' screens per site then their commitment to the megaplex was clear. In 1996: Regal Cinemas had 7.8 screens per site whilst in 2000 it was 10.2; for AMC, the figures were 7.6 and 13.7 respectively and in the case of Cinemark they were 8.2 and 11.3 respectively.[39] There were 23,814 screens nationwide in 1990; by 2000, there were 37,396, which meant that in ten years, 13,582 screens were added.[40]

By the start of the 2000s, AMC was by far the largest operator of megaplexes, with 13 30-screen and 33 24-screen complexes.[41] Nevertheless, Dodona observed that "[d]espite recent over-investment by the industry, the exhibition business in the United States is not a strikingly modern one. Nearly a third of theatres have only one screen and there are still substantial numbers of miniplexes with two to seven screens."[42] Perhaps emblematic of the position faced by several exhibitors was Carmike, which, despite becoming one of the largest circuits in the USA by number of sites, was felt by many in the industry to be particularly vulnerable, since its screen count was much lower than its rivals. In 1998, Carmike had 5.9 screens per site, which was the lowest of the 'Big Five' when compared with 12.3 in the case of AMC and 9.6 in the case of Regal.[43] Many of Carmike's sites were older and lacked the stadium seating that was a feature of megaplexes and many of the later generation of multiplexes. From 1998 onwards, Carmike set about closing many sites, refurbishing others to include stadium seating, and building new cinemas with up to 20 screens. It continued to mitigate its ageing number of sites by having a monopoly position in more than 60 per cent of its markets, many of which were smaller cities in more than 35 US states.[44] By 2000, Carmike was the second largest circuit in the USA; however, the company's expansion, like that of many of its competitors, had meant taking on sizeable amounts of debt. In August 2000, Carmike filed for Chapter 11 protection in order to restructure its business, owing some $653 million against its $841 million in assets.[45] It emerged from bankruptcy in 2002 having cut its circuit from 2848 screens to 2300, with its number of sites down from 500 in 1997 to 308 by 2002.[46]

## CONCESSIONS AND ANCILLARY ENTERTAINMENT—"WHERE ENTERTAINMENT BEGINS"

The commodification of cinema engendered by the multiplex and megaplex found its most emblematic expression in the entrance lobby: "where entertainment begins."[47] Here increasingly was a space for the consumption of not just food or concessions but other kinds of associated goods and services designed to boost revenue—so-called ancillary entertainment—which steadily assumed a greater financial importance in the profitability of the cinema itself. In 1951, sales of 'candy' accounted for 20 per cent of the revenue of a cinema and the concession stand was often referred to as the "second box office" by exhibitors.[48] With the advent of the new two- to six-screen cinemas in the 1960s and 1970s, the placement of the concession stand within the new shared lobby was considered at the architectural design stage. The schemes were to consider the movement of patrons in and out of the auditoria, and the time available, especially as intermissions were steadily phased out in favour of increased number of screenings per day. In the first instance, the smaller auditoria were seen as encouraging sales since with fewer seats per row access to the aisles was easier for patrons to get up and go to the concessions stand whilst the film was showing. The larger lobbies also began to accommodate bigger and more sophisticated concession stands, designed to increase sales. Showcase's forerunner Redstone Theatres incorporated circular stands—the 'Circle of Profit'—in their new multiscreen cinemas which were designed to serve up to 60 patrons at a time with 75 next in line behind them for their four standardised items.[49] In 1986, a major study of sales, sponsored by Nestlé Foods Corporation, concluded that 53 per cent of patrons visited the concession stand and 33 per cent purchased something.[50] The study concluded that it was important to attract patrons when they entered the lobby, since the numbers of people leaving the auditoria during performance had declined since the 1960s and 1970s; whilst the cinemas with the fastest transaction times also had the highest concessions per capita.[51]

In the context of an analysis of the cinema business that suggested that exhibitors generated only 1.5 per cent profit on admission prices,[52] concessions were increasingly important and by 1991, some cinemas were reporting that sales accounted for 90 per cent of profits, with profit margins often as high as 900 per cent in the case of soft drinks.[53] Indeed, as Bruce Proctor, a manufacturer of concessions equipment, pointedly observed: "[t]here's a school of thought which says we're not strictly in the film business—that we're in the food business on at least an equal footing."[54] Indeed, whilst mutually supportive, the businesses of selling tickets and selling concessions do have different objectives, with the selling of concessions essentially a retail business with the objective of maximising spend per head. The megaplex's size and number of screens meant that the concession stand underwent a dramatic change in size and design. Instead of the traditional "walk-up" stand, exhibitors installed so-called pass through counters, which speeded up transactions and increased the range of products.[55] By 2000, attendance was 1.42 billion annually whilst the average concession spend per cinemagoer was $1.53, which meant that concession revenues amounted to $2.17 billion.[56] *Screen Digest* estimated that concessions accounted for 25 per cent of revenues and, there-fore, 50 per cent of profits.[57] By 2000, it was also apparent that any decrease in attendance disproportionately affected overall profits, with the fortunes of the cinema tied closely to the summer months and the supply of blockbuster films.[58] It is perhaps no surprise that the term 'popcorn movie' came to be associated with the summer event or blockbuster films since it these which, according to Salisch, "typically have mass appeal to the audience demographic with the largest disposable income and free time—the pre-teen and teenage offspring of baby boomers—who are attracted by the communal nature of the moviegoing experience and who thrive on snack foods."[59]

Though concessions sales were important to all circuits, Cineplex Odeon focussed relentlessly on this area as a way of boosting revenue, and in 1992 launched what it called "Project Popcorn": a plan to increase the number of patrons who purchased as well as increasing the amount each spent on concessions, with the overall aim of boosting income by $40 mil-lion a year over the following three years.[60] Cineplex Odeon had been keen to focus on concessions when it opened its megaplexes in the 1980s, with Garth Drabinsky negotiating deals with major food and drinks sup-pliers for bulk purchasing and increasing profits partly by employing "inexperienced workers at low wages—like a fast food outlet."[61]

During the 1980s, should audiences manage to negotiate the ever-expanding concession stand they were often presented with the new inno-vation of the video game that began filling in the previously "dead space" in the lobby. In 1979, AMC began installing machines and by 1981 had them in 80 per cent of their 135 cinemas whilst all subsequent develop-ments had them included from the outset.[62] Commenting on the installa-tion of games machines in cinemas, Philip Lowe of the National Association of Concessionaires observed that "[t]heater owners are constantly looking at other ways of maximising their income…It's a way of offering [custom-ers] a total entertainment experience, or stated another way, it's a way of maximising a theater owner's revenue."[63] The multiplex circuits were con-scious of the fact that the video games appealed to the same audience demographic as the cinema but that video arcades had become less attrac-tive to shopping centre managers and municipal authorities; which meant significant commercial opportunities.[64]

In 1996, Cinemark opened a 20-screen megaplex in Plano, Texas, that was indicative of a new trend in megaplex cinemas: that of the 'total entertainment centre,' the 'new entertainment concept,' or, as J. Wayne

Anderson from R/C Theatres put it: "the megaplex is a total entertainment complex which will consist of things other than motion pictures … A megaplex allows more people to do more things in one place and to do so with more members of their peer group or family."[65] Here, Anderson was talking about an increasing trend towards joint ventures with other media-based and electronics companies, which involved the development of new large formats such as IMAX and 3D IMAX, video games, and virtual reality technology. Sega GameWorks (SGW)—a joint venture between MCA Inc., DreamWorks SKG, and Sega in 1995—was emblematic of this new trend. SGW partnered United Artists in opening its Starport entertainment centres in several of its largest sites and Cineplex Odeon in its new Cinescape centres. Cineplex Odeon's Howard Lichtman identified a synergy between its Cinescape and cinema patrons adding that "[i]t is the moviegoer who is going to Cinescape and it is the Cinescape customer who is going to the movies."[66]

Underpinning the attraction of the megaplex to exhibitors was the promise that as the audience appeared to be more fragmented these sites could attract the family audience with a broad range of cinemagoer's tastes, and that a visit to the megaplex would be an "experience." At its Tinseltown USA complexes, Cinemark included pizzerias, coffee bars, and a video games arcade. Additionally, several included an IMAX 3D installation. Valerie Baty, the manager at Cinemark's Tinseltown USA in Plano, drew attention to the complex's "eye-catching" design as signalling an emphasis on a "festive fun atmosphere."[67] According to Brill, the megaplex was "both a harbinger of film exhibition's future and a reaction to changing priorities in how people seek out-of-home entertainment."[68]

In 1994, *Variety* ran the headline "Here Come the Megaplexes" describing the major exhibition circuits as embarked on a "building binge" not simply of theatres but of entertainment "destinations."[69] "Rather than just tacking a theater on the end of a strip mall," intoned Michael Campbell, CEO of Regal Cinemas, "our new theaters will be destinations in themselves."[70] Marcus Theatres' Bruce Olson summed up the strategy for their megaplexes:

[o]ur goal is to create the ultimate in regional entertainment destinations, not only with 20 or more screens, but we're also attaching 50,000 square feet of family entertainment centers. Now the moviegoer can go to this movie complex and know that he or she can see anything that is in release, but have full access to restaurants, attractions, games and other high-tech elements.[71]

However, the entertainment centre concept was not an unqualified success and neither was it risk free. Regal, one of the top five circuits, was also building a large number of megaplexes across the country, as well as its first FunScape in Chesapeake, Virginia, in 1995. This was, the company claimed, the industry's first indoor cinema and entertainment complex,[72] though by 1999 six of the seven built were up for sale or leased, with some doubt as to whether the $6–10 million spent building each one would be recouped.[73]

## EXPANSION, CONSOLIDATION, AND DEBT—"AT SOME POINT THE MADNESS HAS TO STOP"

Such was the capricious nature of the exhibition sector as the new millennium dawned that the optimism of the market and the re-entry of the studios into the sector was almost entirely reversed amidst a plethora of exhibitors who declared Chapter 11 bankruptcy and a further spate of mergers and acquisitions, which resulted in the consolidation of the industry. One of the defining characteristics of the exhibition industry in the late 1990s and early 2000s was its volatility, notwithstanding the fact that between 1995 and 2000 more than 200 million more people went to the cinema and box-office receipts grew by $2.4 billion.[74] Nevertheless, in 2001, according to *Screen Digest*, half of all screens in the USA and Canada (20,000) were operated by companies under Chapter 11 bankruptcy protection.[75] In part, analysts argued, the reason was that the construction of ever-bigger sites by the large exhibition circuits meant that some one third of the 37,000 screens in the USA were surplus to requirements, many of which were in the very same circuits' older venues.[76] As Dodona argued, "for exhibition to return to an even keel, perhaps a quarter of all screens need to close to take the total back under 30,000."[77] Even several of the of the largest operators were not immune, with United Artists Theater Corporation (UATC), Loews Cineplex, Carmike, and General Cinema Corporation (GCC) amongst 11 exhibitors declaring Chapter 11 bankruptcy in the space of two years between 1998 and 2000.

During the late 1990s, there was a frantic period of consolidation within the exhibition industry in the USA. In 1997, UATC, by then the second largest exhibitor behind Carmike Cinemas, itself acquired by AMC in 2016, was the subject of an attempted takeover by a Texas investment company called Hicks, Muse, Tate & Furst though they eventually called off the purchase apparently not prepared to pay the asking price of $850 million.[78] Though UATC continued to be one of the largest circuits in the USA, the company's declining share of the box-office between 1995 and 2000, allied to its decline in net income and ageing, smaller cinemas, meant that it was less attractive as an acquisition. Like many large circuits in the latter part of the 1990s it increasingly lost money, due in no small part to what Dodona argued was "the widespread view that management buy-outs of ageing circuits cannot succeed due to the burden of financing both the purchase price and the investment required for upgrading."[79] In 2000, UATC filed for bankruptcy under a Chapter 11 petition in order to restructure its debts of $667 million. Kurt Hall, United Artists' president and chief executive, said that United Artists' internal cost-cutting plan had failed to "mitigate the impact of the unprecedented level of predatory [expansion] that has occurred in our industry over the past three years."[80] In 2001, the financier Philip Anschutz—United Artists' largest bondholder and creditor—acquired a controlling stake with 55 per cent of the shares.

Having reduced debt and increased revenue, Cineplex Odeon was in the advantageous position of having the majority of its sites in major markets: 85 per cent of its US sites were in the top 15 markets and in Canada the figure was 75 per cent.[81] Having sold some sites to competitors such as Carmike and entering into what the company called "controlled growth," by 1995 Cineplex Odeon was still one of the 'Big Five' with 329 sites and 1535 screens.[82] In May 1998, Cineplex Odeon merged with Loews Theatre Group, itself a part of Sony, to form Loews Cineplex Entertainment (LCE), becoming the second largest exhibitor in the world with 423 sites and 2881 screens by the end of that year including sites and or interests in exhibition, in Spain, South Korea, and Germany.[83] The company was vertically integrated in that it was 51.1 per cent owned by Sony Pictures Entertainment and 26 per cent by Universal. Nevertheless, by 2001 LCE, like many other companies that continued to wrestle with high rates of debt as a result of over-building, filed for bankruptcy under a Chapter 11 petition to protect it from creditors. Citing poor film product in the previous year as the main reason for their troubles, many argued that it was in fact LCE's poor management that was the issue.

One analyst argued that the main cause was "the unbridled expansion and cannibalization of the existing theaters" in the context of an industry that was building more screens than there existed demand for—about 9000 more than the existing movie audience could support.[84] John Krier, president of the Exhibitor Relations Board, argued pointedly that "at some point the madness has to stop."[85] LCE sought Chapter 11 bankruptcy just one month after Regal, then the largest circuit in the USA, and six months after UATC did so. LCE's saviours were in part the LA-based private investment group Oaktree Capital, which along with the Onex Corporation—an electronics and manufacturing company and then Canada's fourth largest company—acquired LCE. In 2003, the new owners reversed the merger between Loews and Cineplex Odeon and merged the latter with the Galaxy circuit to form Cineplex Galaxy Limited Partnership. Despite the closure of several hundred screens as a result of filing for Chapter 11 protection and losing the Loew's sites, Cineplex Galaxy was still one of the five biggest circuits in North America.[86] However, the break-up of LCE saw the company's former Cineplex Odeon cinemas in the USA disposed of as the focus shifted to its Canadian operation, which was consolidated with the purchase of its largest Canadian rival, Famous Players, in 2005. The newly named Cineplex Entertainment then controlled 64 per cent of the Canadian box-office[87] and in 2017 Cineplex Entertainment was easily Canada's largest circuit and the fourth largest in North America with 1683 screens in 165 sites.[88]

Like many of the national circuits that came to dominate exhibition in the 1990s, Regal Cinemas expanded through a process of merger and acquisition, and construction. The company was founded by Michael Campbell and incorporated in 1989. Campbell had owned a small 150-screen circuit called Premiere Theatres, based in Knoxville Tennessee, but sold them a short while before starting Regal and purchasing its first cinema in Florida in 1990. Regal then began to grow into a national circuit through the purchase of cinemas from smaller circuits, such as Litchfield Luxury Theatres and Martin Theatres in 1991, along with sites from established larger circuits such as United Artists. Regal also began acquiring larger regional circuits such as Virginia's National Theaters in 1994, Georgia State Theaters, and Krikorian Theaters, in California in 1996. In that year, the company had more than 1000 screens and in 1997 its acquisition of Magic Cinemas and merger with Cobb Cinemas made it the third largest circuit in the USA. Regal Cinema were no strangers to the consolidation fever of the late 1990s, and in 1998 the company was jointly acquired by Texas investment company Hicks, Muse, Tate & Furst along with another called Kohlberg Kravis Roberts & Co. (KRR) for approximately $1 billion, with the intention of merging the two circuits with Act III Cinemas, itself purchased by KRR in 1997.

In ten years, Regal had become the world's largest cinema circuit, with 403 sites and 3628 screens in more than 30 states in 1999.[89] The success of the company was based in part on its careful market assessments and research into catchments, avoiding direct competition wherever it could and according to Dodona could "claim to be the sole exhibitor in 74% of the film zones where it was present."[90] However, more than half of Regal's circuit were smaller, older sites which had been acquired from other circuits, with new-build sites tending to be larger, more modern multiplexes and megaplexes. Like many of their competitors, Regal's acquisition policy was based in large part on borrowing and in 2001 they also filed for Chapter 11 bankruptcy in order to reschedule their debts, which amounted to $2.29 billion.[91] In the same year as he took control of UATC in similar circumstances, financier Philip Anschutz acquired a controlling interest in Regal Cinemas and the small Southern California–based circuit Edwards, also in Chapter 11 bankruptcy. This meant that the Regal Entertainment Group became the largest circuit in the USA with 6124 screens and 563 sites in 2003,[92] a position it retained until mid-2016, when it had 7310 screens in 564 sites.[93]

In the 1990s, GCC increasingly looked to grow by building ever-larger multiplexes and megaplexes with 14–20 screens, and relinquishing smaller sites. This meant that the number of screens declined in the latter 1990s from 1180 in 1995 to 828 in 2000.[94] The contraction of GCC and the decline in screens to 823 were in part a result of the company's filing for Chapter 11 bankruptcy in October 2000. Despite getting in on the multiplex and megaplex construction boom, GCC had suffered from competition in many areas, not least Texas, and had retrenched its operations in the northeast and Midwest of the country. It sought bankruptcy in order to protect itself from creditors and restructure the company, citing in part what it called "a decline in patronage and profitability … due to the buildup of megaplex theatres over the last several years."[95] In a subsequent auction of the company, GCC was acquired by AMC in 2001 for approximately $167 million.

Throughout the 1990s and into the 2000s, AMC also continued to grow rapidly. In 1996, the company was the second largest circuit after Regal, with 1721 screens in 227 sites.[96] Their acquisition of GCC meant that they assumed control of an additional 66 sites and 621 screens. In August 2012, Dalian Wanda, China's largest property developer, purchased an 80 per cent controlling interest in AMC for $2.6 billion and in March 2016 they purchased rival exhibitor Carmike for $1.1 billion to make AMC the largest cinema circuit in the USA.[97] The deal was approved by the US Department of Justice, subject to AMC divesting itself of sites in 15 geographical areas where both circuits competed directly. Initially, Carmike continued to operate as a wholly owned subsidiary of AMC but less than a year later AMC began 'retiring' the Carmike name and rebranding the cinemas as AMC, a process that was completed by April 2017.[98]

The takeover by Dalian Wanda, headed by Wang Jianlin, was highly significant in that the company had sought to become a globally dominant force in exhibition. Indeed, Wang Jianlin professed that he wanted to own 20 per cent of the world's screens primarily in an effort to strengthen his negotiating position with the main film distributors, notably the major Hollywood studios. Moreover, Dalian Wanda's exhibition interests are part of a growing, vertically integrated film empire with the acquisition of Hollywood production company Legendary Pictures, US-based entertainment marketing company Propaganda Gem, as well as the building of the world's largest production studio in China.[99] The purchase of AMC was followed in June 2015 by the purchase of Australian exhibitor Hoyts, in July 2016, by the purchase of Odeon-UCI, Europe's largest exhibitor, and in January 2017 by the purchase of the Nordic Cinema Group, the largest circuit in Scandinavia and the Baltic states.[100] The latter acquisition was subject to approval from the European Union, which was granted subsequently in March 2017.

## DIGITAL CINEMA—"AN EVOLUTION, NOT A REVOLUTION"

In 1999, 20th Century Fox released *Star Wars: Episode I— The Phantom Menace* on digital, though it was shown in only four cinemas in June of that year. Digital cinema was trumpeted by many, including Lucas, as the next revolution, though as Belton pointed out, whilst production and post-production were increasingly dependent on digital technologies "this reliance on the digital domain was relatively invisible to the average moviegoer."[101] Belton cited *Phantom Menace* producer Rick McCallum, who suggested that the premiere was "a milestone in cinematic history" and "that like the introduction of sound and color, these digital screenings represent the beginning of a new era in film presentation."[102] Despite the hyperbole in some of the press, however, the number of cinemas equipped for digital projection remained small—88 screens in 2003.[103] What growth there was effectively stopped in 2003, when the seven major studios formed a consortium called the Digital Cinema Initiative (DCI) in order to produce what was called the *Digital Cinema System Specification V1.0*, published in July 2005. In consultation with the Society of Motion Picture and Television Engineers (SMPTE), this set out industry-wide standards for the mastering, distribution, and theatrical playback of digital content, with the studios forecasting that "the introduction of Digital Cinema has the potential for providing real benefits to theater audiences, theater owners, filmmakers and distributors."[104] The publication of the DCI's specification defined the system requirements for projection equipment—digital light projectors (DLPs)—and the "performance characteristics used to display the image on the screen" and the standards for encryption and copyright protection.[105]

Prior to the publication of the DCI's standards there was a kind of "phoney war" when whatever small growth there was in the installation of DLPs was tempered by a vigorous debate between producers and exhibitors about the relative benefits of the technology and anxieties about cost.[106] For the major circuits the issue was not technological but financial as although the potential savings to be made were to the distributors the cost of installation fell largely on the exhibitors. In 2000, Steven Morley, head of Qualcomm's digital media division, argued that since nothing was "broken" about the way film was used then the only way digital cinema would have an impact would be if it added "value or capability" and made money for the industry.[107] Morley added that the industry would generate profit by increasing revenues—selling more tickets per screen or developing new revenue streams such as alternative content/live events—or decreasing costs—eliminating film duplication and distribution expenses.[108] Thus, what would happen would be "an evolution, not a revolution."[109]

The possibility for cutting costs in the area of distribution was an attraction to the major Hollywood studios, especially since most of them operated distribution arms as well. As Katz et al. observed: "[i]t is the studios that have the most to gain in the short-term, as they currently bear the brunt of the print and distribution costs, an estimated $1 billion per year in the U.S. The exhibitors thus have limited incentive in the short-term to shell out $150–$200,000 per screen to upgrade their theaters."[110] Thus, diffusion of the technology was more happenstance in the early days of its development. Hancock summed up the situation concisely:

[t]he digital cinema market is now entering a deployment phase...The key players now have a chance to position themselves for the cinema market of the future. It is striking that several territories are moving ahead on their

own paths, not waiting for the main industry drivers: the US studios. However, as many of these pioneering systems are low-to mid range, high-end deployments will not begin in earnest until the studios themselves enter the global fray. Digital distribution and exhibition is about to replace the multiplex as the next battle ground for the re-generation of cinema economics.[111]

In 2005, the exhibition industry seemed poised to convert to digital projection subject to some agreement being established between them and the distributors or projector manufacturers as how to finance conversion. In 2007, Carmike, still one of the 'Big Five' albeit in a distant fourth place behind companies like Regal, was the circuit that embraced the emerging digital projection technologies most enthusiastically. Carmike had installed digital projectors in 1340 of its screens in 2007, compared to Regal's 22 and AMC's 15.[112] Carmike's decision to instal the new Christie projectors was part of a deal with the digital integrator Access IT and reflected the positioning of their cinemas in smaller cities with a consequent lower frequency of visits per screen compared to their larger, metropolitan-based competitors. For Carmike, the cost of film prints was of greater significance than it was for rivals and digital projection, with the emergence of virtual print fees[113] and the subsidised installation and service costs, offered the possibility of considerable cost savings in the mid-to long-term.[114] By 2010, Carmike's circuit was almost entirely digital, with 2103 out of 2236 screens converted, with its increase in screens in part a result of continued acquisition of smaller cinema circuits.[115]

The rapidity with which the major circuits converted to digital projection was impressive. In 2006, there were already 2014 digital screens but by 2010 the total was 15,774, half of which were equipped for 3D, with speculation too that "it seems likely that 35mm projection will be becoming a curiosity around the end of 2013."[116] By 2013, only 8.74 per cent of screens were non-digital.[117] In January 2014, Paramount became the first major studio to announce that it was ceasing distributing films on 35mm, with its release of *The Wolf of Wall Street* (2013) being the first major title available digitally only.

Though much of the impetus to convert was due to the demands of distributors, with the technology essentially "invisible" to cinemagoers; the development of digital 3D was seen as a technology that could be marketed by exhibitors as a novelty for which they could charge more. Though 3D emerged initially in the 1950s, its attractions were short-lived and diffusion limited. However, the new digital 3D system, first show-cased to large audiences in December 2009 with *Avatar*, promised to be more attractive and durable; whilst the increased admission fee could help to pay for the installation of the new technology. There was also a sugges-tion that 3D was part of a successful strategy to force exhibitors to fund and instal digital projection.[118] In 2010, *Avatar, Alice in Wonderland,* and *Toy Story 3* took a combined total of $1.22 billion at the US box-office, with some 20.8 per cent of that accounted for by 3D.[119] In 2008, there were 1427 digital 3D screens in US cinemas but by 2010 there were 7837.[120] James Cameron, via his company Lightstorm Entrainment, was an early evangelist for the format and in 2005, his partner Jon Landau was of the view that "3D is one of those things that people will come out of their homes in droves to see. From the big-scale movies to the small dra-mas."[121] As Belton observed, "[t]he argument was that digital 3D finally offered audiences something they could not get at home or in conven-tional, non-digital, movie theaters."[122] However, 2010 would seem to have been the high watermark for 3D in terms of box-office since although installations continued apace—in 2013 there were 14,483 3D screens—the proportion of the box-office accounted for by 3D films had declined to 16.5 per cent.[123] By 2017, according to the Motion Picture Association of America (MPAA), the trend was resolutely down with 3D films account-ing for 12 per cent of the combined US/Canadian box-office.[124] For Belton, the problems associated with 3D for exhibitors were in part because of a "revolt" against the surcharge for tickets and accusations of poor product from the studios.[125] Whilst the novelty stage of 3D might well have boosted the rollout of digital projection its diffusion depended on it ceasing "calling attention to itself."[126]

## Now It Is the 'Big Three'

The increasing concentration of screens in multiplexes proceeded through-out the 2000s so that by 2017, 89 per cent screens were in multiplexes (sites with eight screens or more).[127] In terms of ownership, the largest three cinema circuits in the USA—AMC, Regal Entertainment, and Cinemark, which constituted a new 'Big Three'—owned 1548 sites, which accounted for 49.7 per cent of total screens and with an average of 12.9 screens per site.[128] Though all of these circuits had undergone signifi-cant restructuring and mergers, or were the result of mergers, AMC and Regal had come to dominate the market in the period from the 1970s onwards, joined by Cinemark in the 1980s. Moreover, all had built their supremacy of the market on their development of the multiscreen cinema, particularly the multiplex and megaplex. This domination of screens and sites meant that in 2017 these three circuits accounted for 72.8 per cent of all admissions and 57.5 per cent of the box-office.[129] Significantly, though, the largest two of those circuits were foreign-owned, since Regal were purchased by British circuit Cineworld in March 2018 for $3.4 billion, making Cineworld the second largest circuit in the world by number of screens (9500); second only to AMC, owned by Dalian Wanda, which had 11,000. According to the National Association of Theatre Owners (NATO), the industry is buoyant and progressive:

> [a]t the same time, the industry has not held still, investing billions of dollars to completely change our core projection technology, creating opportunities for new technologies like 3D, high dynamic range, high frame rates and immersive sound. Theaters now offer Premium Large Format screens, restaurant and alcohol service, reserved ticketing, loyalty programs, and ticketing apps, all while maintaining an average ticket price that, adjusted for inflation, is lower than the average ticket price in 1976.[130]

This optimism seemed confirmed by the news that in 2018 the box-office in the USA passed 2016's record of $11.382 billion with a week to spare, whilst attendance was up 4 per cent over the previous year.[131] This was especially significant as in 2017 the box-office had not maintained the increase during 2016, but more alarmingly, admissions had dropped 6 per cent to 1.23 billion, the lowest in 23 years.[132]

## The Home of the Multiplex and the 'Retail Apocalypse'

When Stanley Durwood opened the Parkway Twin in 1963 the Ward Parkway Mall was one of more than 3800 such shopping centres in the USA and, as we have seen, the relationship between the multiplex and the mall has been a symbiotic one. Nevertheless, by 2017 there were approximately 1100 malls in the USA, greatly reduced from the 36,000 open in 1990, with a quarter of those left at risk of closing within five years.[133] The term that has entered into the popular lexicon to define this trend is the 'retail apocalypse': a 'tipping point,' in which the restructuring of retail shopping is both permanent and something physical retailers cannot recover from.[134] Despite an intuitive desire to simply blame the rise of Internet retailers such as Amazon, the reasons are varied and complex including: the evident changes in the way people shop and the attraction of online retailing in an age in which people afford less time to shopping as leisure; the large amounts of debt incurred on the part of mall owners and large retailers like department stores, with continued restructuring of these debts increasingly difficult since the 2008 crash; the decline in real wages particularly since the crash in 2008 too; an oversupply of malls; and, perhaps most dramatically, the diminished place of malls in the day-to-day lives of young people. Many younger Americans, according to filmmaker Dan Bell, "look at malls in an antiquated way … they see it as, 'That was my parents' thing, and it's not my thing.'"[135]

Nonetheless, there are still malls and many of those are looking again at their design and layout, and seeking to re-present the "shopping experience." The emphasis is shifting towards restaurants and eating, leisure facilities such as gyms, different kinds of retailing, and, significantly, the continued place of the cinema, often newly opened or converted as luxury venues in malls being repositioned as upmarket spaces. This has been helped, as far as the circuits are concerned, by the relaxation of regulations regarding sale of alcohol in cinemas in the majority of states, which have also proven to be extremely profitable. As George Patterson from AMC argued: "It is the fastest-growing amenity in our industry. Dollar for dollar, alcohol doesn't cannibalize candy, soda or popcorn. If I put a bar in, I can almost guarantee my food and beverage revenue will go up. Over 70 percent of our guests are over the age of 21."[136] At the formally declining Staten Island Mall, which was being renovated, for instance, AMC is opening a new 45,000-square-foot, 11-screen multiplex equipped with over 1000 reclining seats and billed as one of the circuits 'Dine-in' cinemas.

## THE MULTIPLEX GOES INTERNATIONAL—'A SOFT REVOLUTION'

By the mid-1980s, confident of the appeal of Hollywood films in Europe and parts of Asia, including Australia, and the poor state of many cinemas in comparison to those in the USA, US companies felt that the vast capital outlay necessary for expansion into overseas exhibition could be recouped relatively quickly. In view of their assessment of the technological backwardness of many of the world's cinemas, many company executives felt that they were on a mission to change exhibition abroad. Indeed, Warner

Bros. executive Salah Hassanein argued that his company's overseas expansion was a "reaction to the lethargy that has kept theaters abroad far behind their North American counterpart technologically," adding that "[t]here has needed to be a change for a long time, and we are responding to that."[137]

Like his counterparts in other cinema companies, Hassanein might have been conscious that the overseas exhibition market could become twice or even three times the size of the $5 billion-per-year US market by the end of the century. There were two other factors that encouraged market entry overseas: the over-saturation in the US market and the desire of the major film distributors to increase their revenues from European markets.[138] In 1965, there had been 13,000 screens in the USA which rose to 24,000 in 1988; however, Dodona argued that the last 6000 of these screens accounted for only a 6 per cent rise in admissions.[139] This acted as a "push factor" for the large circuits to find these new markets. The reasoning on the part of studios was that multiplexing abroad would influence the growth of film distribution and, in the words of an executive from UK-based United International Pictures, "the faster we can get up to the level of having a cinema infrastructure analogous to that of the U.S., the faster we can regenerate the health of the industry in its entirety."[140] Moreover, another industry executive observed that this dominance would lead to a "soft revolution" in which cinemagoers would be won over by better concessions (drinks and food), better seating, and better auditoria.[141] Driving the look to overseas markets was also a sense amongst exhibitors and distributors, especially prior to the development of the megaplex, that the US market was 'over-screened.' Therefore, it would be to Western Europe that the major circuits would look first and Britain in particular.

## Notes

1. John Hazelton, "Drowning in a Sea of Screens," *Screen International* 734 (9 December, 1989), 17.
2. "Mixed Bag for Worldwide Cinema Last Year," *Screen Digest* (February, 2003), 61.
3. Dodona, *Cinema Industry Research: USA & Canada* (Leicester: Dodona Research, 2014), 1.
4. In 2001, AMC closed eight screens on the lower level due to declining ticket sales, with the fourteen screens on the upper level remaining open.

5. Some in the trade press referred to these large 14+ screen complexes as "Ultraplexes" though only the regional circuit, CinemaStar Luxury Theatres, chose to call their new 14-screen megaplex, opened in 1996 in Riverside, California, the "Ultraplex 14."

6. Dodona. *Moviegoing* (Leicester: Dodona Research, 1999).

7. Kevin Lally and Ed Kelleher, "Special Report: Megaplex Mania," *Film Journal International* 99, no. 7 (1 August, 1996), 3.

8. "AMC Opens Largest Theatre Complex in U.S.," *The Film Journal* 98, no. 5 (1 June, 1995), 22.

9. AMC Entertainment Inc., *1995 Annual Report* (Kansas City, MO: AMC Entertainment Inc., 1995), 12.

10. Debra Skodack, "Going Grand in Dallas," *Kansas City Star*, 9 June, 1994.

11. "AMC Begins Construction on Largest U.S. Multiplex," *The Film Journal* 97, no. 6 (1 July, 1994), 24.

12. Braden Phillips, "Regal at 10: Stadium Seating Tops List of Theater Design Priorities," *Variety* 374, no. 3 (8 March, 1999), 36.

13. Michael L. Campbell, "The State of our Art: The Best and Worst of Times: The Head of Regal Cinemas, Newly Crowned King of the Exhibition Universe, Gives *BoxOffice's* First Annual State of the Industry Address," *BoxOffice* 135, no. 1 (1 January, 1999), 22–23.

14. AMC Entertainment Inc., *1999 Annual Report* (Kansas City, MO: AMC Entertainment Inc., 1999).

15. Campbell, "The State of our Art," 22.

16. Ibid., 22.

17. Lally and Kelleher, "Special Report: Megaplex Mania," 138.

18. Quoted in Susan Hightower, "AMC's Grand Multiplex," *Kansas City Star*, 18 May, 1995.

19. Andrew Hindes, "AMC at 30—Megaplex Dominance Intact," *Variety* 370, no. 5 (16 March, 1998), 39.

20. Dodona, *Cinema Industry Research: USA & Canada*, 13.

21. Patrick von Sychowski, "Every Multiplex is New Again," *Cinema Technology* 31, no. 1 (March, 2018), 22.

22. Ibid.

23. Lally and Kelleher, "Special Report: Megaplex Mania," 20.

24. See Michael D. Beyard and W. Paul. O'Mara, *Shopping Center Development Handbook.* Third Edition (New York: Urban Land Institute, 1999).

25. Cited in Myron Meisel, "Entertainment Centers: Edwards Launches 'the Big One'," *Film Journal International* 99, no. 3 (1 March, 1996), 44.

26. Robert La Franco, "My Megaplex is Bigger than Your Megaplex," *Forbes*, 24 February, 1997. https://www.forbes.com/forbes/1997/0224/590 4050a.html#5d8394277a4d (accessed 14 October, 2018).

27. Peter Henné, "Dueling Megaplexes," *Film Journal International* 100, no. 5 (1 June, 1997), 10, 82.
28. Lally and Kelleher, "Special Report: Megaplex Mania," 20.
29. Dodona, *Moviegoing* (Leicester: Dodona Research, 1999).
30. Ross Melnick and Andreas Fuchs, *Cinema Treasures: A New Look at Classic Movie Theaters* (St. Paul: MBI, 2004), 180.
31. Leonard Klady, "Megas Lack Plexuality," *Variety* 366, no. 5 (3 March, 1997), 86.
32. Lally and Kelleher, "Special Report: Megaplex Mania," 24.
33. Klady, "Megas Lack Plexuality."
34. Ralf Ludemann, "National Amusement at 60: On with the Show," *Screen International* 1085 (22 November, 1996), 16.
35. Dodona, *Moviegoing Issue Four* (Leicester: Dodona Research, December, 2001), 8.
36. Lally and Kelleher, "Special Report: Megaplex Mania," 3.
37. Roger Smith, "The Street Mulls Megaplex Mania," *Variety* 377, no. 5 (13 December, 1999), 6.
38. Dodona, *Moviegoing Issue Two* (Leicester: Dodona Research, April, 2001), 6.
39. Dodona, *Moviegoing 4* (Leicester: Dodona Research, 2004), 32. and "Worldwide Cinema: Poor Product Fails the Multiplex Boom," *Screen Digest* (September, 2000), 283.
40. Dodona, *Moviegoing Issue Four*, 5.
41. AMC, *2001 Annual Report* (Kansas City, MO: AMC Entertainment Inc., 2001)
42. Ibid., 5.
43. Dodona, *Moviegoing 4*, 36.
44. Brown Colin, "Carmike Continues Expansion," **Screen International** 1111 (6 June, 1997), 4.
45. Francesca Dinglasan, "Carmike, Edwards file for Bankruptcy; Regal Next?," *BoxOffice* 136, no. 10 (1 October, 2000), 62.
46. Dodona, *Moviegoing 4*, 39.
47. Carolyn Heinze, "Today's Lobby: Where Entertainment Begins," *Film Journal International* 103, no. 11 (1 November, 2000), 80.
48. Maggie Valentine, *The Show Starts on the Sidewalk: An Architectural History of the Movie Theatre, Starring S. Charles Lee* (New Haven and London: Yale University Press, 1994), 182.
49. Gary Burch, "The Psychology of Concession Sales; A Look at the Subtlety of Selling," *BoxOffice* 114, no. 24 (19 March, 1979), a16.
50. "Candy Bolsters Concession Sales," *BoxOffice* 122, no. 3 (1 March, 1986), 68.
51. Ibid., 70

52. John Larmett, Elias Savada, and Frederic Shwartz Jr, *Analysis and Conclusions of the Washington Task Force on the Motion Picture Industry* (Washington, DC: Government Printing Office, March 1978), 16. http://www.mecfilms.com/mid/task.htm (accessed 12 November, 2017).

53. Charles Fleming, "Snackbar Slowdown Bitter Pill for Exhibs," *Variety* 342, no. 1 (21 January, 1991), 3.

54. Bruce Proctor, "Building for the '90s Includes Better Concession Facilities," *The Film Journal* 96, no. 2 (1 March, 1993), 44.

55. "Cinema Concessions: An Essential Luxury," *Screen Digest* (July 2002), 219.

56. Ibid., 216.

57. Ibid., 213.

58. Anita Watts, "Attendance and Concessions: We Need Movies! Snack Corner Part II," *Film Journal International* 103, no. 11 (1 November, 2000), 42.

59. Wynn. J. Salisch, "The Impact of Programming on Concession Sales," *Film Journal International* 102, no. 11 (1 November, 1999), 60.

60. Karen Murray, "Cineplex Odeon 15th Anniversary—Exhibber Back from the Brink," *Variety* 355, no. 2 (25 July, 1994), 32.

61. Douglas Gomery, "Thinking about Motion Picture Exhibition," *The Velvet Light Trap* 25 (Spring, 1990): 3–11, 6.

62. "Bleep! Zap! Ding! Profits!" *BoxOffice* 117, no. 9 (1 September, 1981), 37.

63. Quoted by Garry Edgerton, "The Multiplex: The Modern American Motion Picture Theater as Message," *Journal of Popular Film and Television* 9, no. 4 (Winter, 1982): 158–65, 161.

64. Dina Lebo, "New Ideas for Exhibition: Arcade Alternatives," *Film Journal International* 99, no. 5 (1 June, 1996), 49, 93.

65. Louis M. Brill, "Megaplex Rising," *BoxOffice* 130, no. 10 (1 November, 1994), 44.

66. Quoted in Pat Kramer, "Let The Games Begin!," *BoxOffice* 133, no. 11 (1 November, 1997), 60.

67. Francesca Dinglasan, "Operating Megaplexes: Megaplexing Issues," *BoxOffice* 135, no. 9 (1 September, 1999), 56.

68. Brill, "Megaplex Rising," 44.

69. Paul Noglows, "Here Come the Megaplexes: Exhibs Usher in 24-screen 'Destinations'," *Variety* 356, no. 4 (22 August, 1994), 1.

70. Ibid., 1.

71. Lally and Kelleher, "Special Report: Megaplex Mania," 3.

72. "Regal At 10: Regal Cinemas Corporate History," *Variety* 374, no. 3 (8 March, 1999), 42.

73. Dodona, *Moviegoing*, 38.
74. Dodona. *Moviegoing Issue One* (Leicester: Dodona Research, 2001), 2.
75. "Global Cinema Exhibition: Dollar Revenues Fail to Keep up with Soaring Admissions," *Screen Digest* (September, 2002), 279.
76. "Worldwide Cinema: Poor Product Fails the Multiplex Boom," 259.
77. Dodona, *Moviegoing Issue Three* (Leicester: Dodona Research, 2001), 4.
78. Colin Brown, "UATG Goes Back on Block," *Screen International* 1147 (27 February, 1998), 6.
79. Dodona, *Moviegoing Issue One*, 14.
80. *Los Angeles Times*, 6 September, 2000.
81. Murray, "Cineplex Odeon 15th Anniversary," 54.
82. Pat Kramer, "1995 Giants of Exhibition: Overview: Land of the Giants," *BoxOffice* 131, no. 12 (1 December, 1995), 28.
83. Dodona, *Moviegoing*, 40.
84. Terry Pristin, "Loew's Cineplex May Seek Bankruptcy Protection," *New York Times*, 24 January, 2004.
85. Quoted by La Franco, "My Megaplex Is Bigger Than Your Megaplex."
86. Dodona, *Moviegoing 4*.
87. Dodona, *Moviegoing 6* (Leicester: Dodona Research, 2007), 8.
88. North American Theater Owners (NATO), *Top 10 U.S. & Canadian Circuits as of July 1 2016* (2017). http://www.natoonline.org/data/top-10-circuits/ (accessed 29 March, 2017).
89. "Regal At 10: Regal Cinemas Corporate History," 38.
90. Dodona, *Moviegoing*, 37.
91. *Los Angeles Times*, 13 October, 2001.
92. Dodona, *Moviegoing 4*, 32.
93. NATO, *Top 10 U.S. & Canadian Circuits as of July 1 2016*.
94. Dodona, *Moviegoing Issue Two*, 12.
95. Dinglasan, "Operating Megaplexes," 140.
96. Dodona, *Moviegoing 4*, 32.
97. Charles Clover, "Dalian Wanda's AMC to Buy Carmike Cinemas for $1.1bn," *Financial Times*, 4 March, 2016.
98. By 2017 AMC had rebranded the whole of its chain into three types: AMC Theatres, which are premium sites with reclining chairs, technologies such as 3D and IMAX; AMC Classic Theatres or "America's Hometown Theatres," which are AMC's older and smaller sites with less traffic, and which feature cheaper entry and concession offers; and AMC Dine-in Theatres, which are marketed as a "restaurant and movie theatre rolled into one" and which feature waiter service to your seat.
99. David Hancock, "Waving the Magic: Wanda," *Cinema Technology* (December, 2016): 14–15, 17.

100. Patrick Frater, "Wanda Expands Global Theater Reach as AMC Pays $929 Million for Nordic Cinema," *Variety* (23 January, 2017). https://variety.com/2017/film/finance/wanda-expands-global-theater-reach-as-amc-pays-929-million-for-nordic-cinema-1201966877/ (accessed 12 July, 2018).
101. John Belton, "Digital Cinema: A False Revolution," *October* 100 (Spring, 2002): 98–114, 103.
102. Ibid., 104.
103. Dodona, *Moviegoing 4*, 18.
104. Digital Cinemas Initiative, *Digital Cinema System Specification V1.0* (Digital Cinema Initiatives, Hollywood, CA, 20 July, 2005), 1.
105. Ibid., 5.
106. Karina Aveyard, *Lure of the Big Screen: Cinema in Rural Australia and the United Kingdom* (Bristol: Intellect, 2015).
107. Steven Morley, "Fitting Digital into Theatres: Merging Technology with Existing Operations," *Film Journal International* 103, no. 3 (1 March, 2000), 76.
108. Ibid., 78.
109. Ibid., 76.
110. Mike Katz, John Frelinghuysen, and Krishan Bhatia, *Digital Cinema: Breaking the Logjam* (McLean, VA: Booz Allen Hamilton Inc., 2002), 2.
111. David Hancock, "Digital Cinema—The End of the Beginning, Press Release," *Screen Digest* (25 May, 2005). http://www.screendigest.com/press/releases/FHAN-6CJJDS/pressRelease.pdf (accessed 18 July, 2006).
112. Dodona, *Moviegoing 6*, 34.
113. The Virtual Print Fee (VPF) model is a means of financing the conversion to digital cinema. A third party pays up front for the equipment, and then recoups the cost of the equipment over time, through payments from distributors (who pay the majority of the cost) and exhibitors.
114. Dodona, *Moviegoing 6*, 34.
115. Dodona, *Moviegoing 10* (Leicester: Dodona Research, 2011), 22.
116. Ibid., 11.
117. MPAA. *Theatrical Market Statistics 2015* (Motion Picture Association of America Inc., 2016), 25. http://www.mpaa.org/wp-content/uploads/2016/04/MPAA-Theatrical-Market-Statistics-2015_Final.pdf (accessed 6 January, 2017).
118. Thomas Elsaesser, "The 'Return' of 3-D: On Some of the Logics and Genealogies of the Image in the Twenty-First Century," *Critical Inquiry* 39 (Winter, 2013): 217–46.
119. Dodona, *Moviegoing 10*, 12.
120. Dodona, *Cinema Industry Research: USA & Canada 2014*, 5

121. Quoted in David M. Halbfinger, "Going Deep for Digital," *New York Times*, 26 September, 2005. http://www.nytimes.com/2005/09/26/business/media/26digital.html?pagewanted=all (accessed 20 November, 2017).

122. John Belton, "Digital 3D Cinema: Digital Cinema's Missing Novelty Phase," *Film History* 24 (2012): 187–95, 191.

123. Dodona, *Cinema Industry Research: USA & Canada*, 5.

124. MPAA, *2017 Theme report* (Motion Picture Association of America Inc., 2018), 15. https://www.mpaa.org/wp-content/uploads/.../MPAA-THEME-Report-2017_Final.pdf (accessed 7 December, 2018).

125. Belton, "Digital 3D Cinema," 192.

126. Ibid., 194.

127. Dodona, *Cinema and Industry Research: USA & Canada* (Leicester: Dodona Research, 2018).

128. Ibid., 7.

129. Ibid., 1, 11–4.

130. "Comments by John Fithian President & CEO National Association of Theatre Owners. CinemaCon State of the Industry," Las Vegas, Nevada, 28 March, 2017. https://www.natoonline.org/wp-content/uploads/2017/03/Fithian-CinemaCon-Speech-2017.pdf (accessed 5 July, 2018).

131. Rebecca Rubin, "Record Breaker! Domestic Box Office Hits New Year High in 2018," *Variety* (24 December, 2018). https://variety.com/2018/film/news/domestic-box-office-record-2018-1203095448/ (accessed 28 December, 2018).

132. Steven Zeitchik, "Is Netflix Killing the Movie Theater? Not so Fast," *The Washington Post*, 24 December, 2018. https://www.washingtonpost.com/business/is-netflix-killing-the-movie-theater-not-so-fast/2018/12/24/7a16dbf8-037a-11e9-8186-4ec26a485713_story.html?noredirect=on&utm_term=.273d927a5d93 (accessed 28 December, 2018).

133. Josh Sanburn, "Why the Death of Malls Is About More Than Shopping," *Time*, 20 July 2017. http://time.com/4865957/death-and-life-shopping-mall/ (accessed 27 December, 2018).

134. Sabrina Helm, Soo Hyun Kim, and Silvia Van Riper, "Navigating the 'Retail Apocalypse': A Framework of Consumer Evaluations of the New Retail Landscape," *Journal of Retailing and Consumer Services* (In Press), https://doi.org/10.1016/j.jretconser.2018.09.015 (accessed 28 December, 2018). See also: Matt Townsend et al., "America's 'Retail Apocalypse': Is Really Just Beginning," *Bloomberg* (8 November, 2017), https://www.bloomberg.com/graphics/2017-retail-debt/ (accessed 23 December, 2018).

135. Quoted in Sanburn, "Why the Death of Malls Is About More Than Shopping."

136. Quoted in Pamela McClintock, "A 'Fifty Shades' Cocktail? How Booze Is Becoming the New Popcorn at Movie Theaters," *The Hollywood Reporter*, 2 February, 2017. https://www.hollywoodreporter.com/news/a-fifty-shades-cocktail-how-booze-is-becoming-new-popcorn-at-movie-theaters-970726 (accessed 27 December, 2018).
137. Ibid., S-3.
138. Dodona, *Cinemagoing* (Leicester: Dodona Research, 1989), 20.
139. Ibid., 20.
140. Michael Williams-Jones quoted in Estep Nagy, "Doing the Continental," *BoxOffice* 131, no. 7 (1 July, 1995), 32.
141. *The Hollywood Reporter* 317, no. 25 (May, 1991), S-4.

# PART II

# The Multiplex Cinema in Britain

# CHAPTER 5

# Exporting the Multiplex to Britain

The mid-1980s mark a crucial moment in cinema exhibition history in Britain. By 1984, the audience had virtually disappeared—declining from 1.6 billion admissions in 1946 to 54 million. Unlike Britain though, as we have seen, the cinema sector in the USA remained dynamic throughout the 1970s and 1980s, with a year-on-year increase in the number of screens and much new cinema building. Writing at the time on the state of cinema in Britain, Roddick characterised the background as "a vanishing audience, a struggling industry and an indifferent, if not hostile state."[1] The parlous state of the exhibition sector in 1984 was examined by *Screen Digest*, which forecast that the decline in audiences in the first few months of the year could translate into the closure of 400 cinemas.[2] The situation looked hopeless. However, 1984 would be the nadir since in 1985 admissions rose for the first time in decades. For many in the industry, and importantly in the USA, the turning point was the designation, amid a wave of publicity and jingoism, of 1985 as British Film Year. Its target was a modest 4 per cent growth, whilst in the event the projected 56 million admissions turned out to be 70.4 million. Finally, in December 1985, in the expanding new town of Milton Keynes, American Multi-Cinema (AMC) opened The Point, Britain's first purpose-built multiplex cinema.

© The Author(s) 2019
S. Hanson, *Screening the World*,
https://doi.org/10.1007/978-3-030-18995-2_5

## The Exhibition Landscape in the Period Prior to 1985—From Palace to Fleapit

In 1984, cinemagoing and cinemas were in a dire state. There were only 660 cinemas,[3] half the number from ten years before, whilst annual admissions stood at 54 million, with 74 per cent of the adult population not attending the cinema at all.[4] This decline in cinema attendance, particularly from the early 1960s, came to be seen as a 'slow death,' as every year produced ever-lower number of cinemagoers and many cinemas closed. Britain was viewed as a market in which the domestic exhibition sector was in terminal decline, whilst at the same time being a market in which films from the USA were both popular and dominant.

Throughout the 1970s, as the decline in cinemas and cinemagoing accelerated one of the responses was the establishment by Prime Minister Harold Wilson, in August 1975, of a working party under the chairmanship of John Terry[5] to report on "the requirements of a viable and prosperous British film industry over the next decade."[6] The working party's focus was largely on film production; however, it was the subsequent follow-up that was significant for the exhibition sector. When Wilson resigned as Prime Minister his successor, James Callaghan, appointed him as chair of an Interim Action Committee, charged with following up on the working party's proposals. In 1977, the recently formed Association of Independent Producers (AIP) convened an All-Industry Committee that bought together some 17 trade bodies under the chairmanship of screenwriter Robert Bolt, which submitted a report to Wilson's committee. Their message, with regard to exhibition, was, according to a report in *Screen International*, stark: "British cinema will die, unless the government is prepared to give aid to film exhibitors."[7] The committee concluded that whilst television was the principal reason for the high rates of cinema closures, there were also additional factors including "unattractive physical conditions" and "insufficiency of choice," which, allied to drastic cost cutting in the industry, had made the condition of Britain's cinemas "often deplorable."[8] However, for representatives of the exhibition industry the answer was to call for the further subdividing of single-screen cinemas, for which government funds should be available in the short term to enable the industry to weather the crisis. In 1981, *Variety* contrasted the slight rise in admissions in 1978 due to a string of Hollywood features with cinemagoers' views on the state of Britain's cinemas which included "discomfort, offhandedness, noise" and "dirty."[9] In short, the article argued, exhibitors' cutbacks in staffing, heating, lighting, and a failure to refurbish meant that the "palace" became a "fleapit."[10]

The comments from the industry and critical observers on the poor state of the cinema experience for cinemagoers would find several echoes as the 1970s gave way to the 1980s, and attendances continued to dwindle. Amongst several surveys of cinemagoers undertaken in this period, many indicate dissatisfaction. In 1982, *Screen International* questioned 100 people on the streets of London, of whom 54 per cent felt that the cinema as "not comfortable" with smoking, poor seating, and lack of space singled out.[11] Three years later, in 1985, London radio station LBC undertook a small survey asking peoples' opinions of cinema in the capital and surrounding area. Of the 88 people who responded 19 per cent said "dingy cinemas" dissuaded them from attending, whilst 20 per cent commented on the poor technical quality of film prints and sound, and 14 per cent did not like the small auditoria created when existing single-screen cinemas were divided up into twin, triple, and quadruple screens.[12] Finally, the survey asked respondents what suggestions they had to make the cinema and cinemagoing more attractive. As a portent of the multiplex cinema, the most popular responses included "advance ticket buying facilities" and "restaurants within the cinema complex."[13] The cinemagoers in this survey were reflecting a general feeling amongst cinemagoers that cinema had lost much of its attraction in the preceding decades. For John Hogarth of independent distributor Enterprise Pictures, the state of Britain's cinemas was to blame for the decline in attendance in the early 1980s and in particular a lack of investment in theatrical refurbishment.[14]

By 1984, a series of trends had pushed cinema in Britain to the brink, including: a contraction of the cinema exhibition infrastructure; a corresponding decline in and fragmentation of the cinema attendance; an uneven distribution of cinema provision, and a lack of imagination and investment on the part of exhibitors.

## A CONTRACTION OF THE CINEMA EXHIBITION INFRASTRUCTURE AND DECLINE IN ATTENDANCE—THE FIRST MULTISCREEN CINEMAS

Though the number of cinemas declined year on year in the post-war period, from the late 1960s in particular, this contraction was offset somewhat by the subdividing of auditoria and the creation of multiscreen cinemas in what had been single-screen sites. So, in 1969 there were 1559 cinemas with 1581 screens, which more than halved to 707 cinemas in 1983 but with 1293 screens.[15] However, the corollary of this multiscreening was a dramatic fall in the size of cinema auditoria, with *Screen Digest* concluding that there had been a "net fall of 75 per cent" in the numbers of cinemas seating 500–1000 since the early 1960s; whilst two thirds of cinemas had fewer than 400 seats.[16] This trend reflected the dramatic decline in the numbers accommodated in Britain's cinemas. Between 1952 and 1964, the total seating capacity of all cinemas halved from 4.2 million to 2.1 million[17] and by 1974 it had more than halved again to 973,000.[18] In 1984, on the eve of the multiplex, seating capacity had shrunk yet further to 459,000.[19] Whilst the 'twinning' and 'tripling' of cinemas undoubtedly helped exhibitors to stress the benefits of greater choice, there was considerable variation in the quality of conversions. In many cinemas, Brown argued, screens had been poorly positioned and the rake on the seating was too shallow for cinemagoers to see over the heads of the people in front.[20] For Sexton "these bedsit auditoria made watching a film purgatorial: tiny screens, awkward seating, over-bright lighting, with shots and screams ringing out from the similar hellhole next door."[21] Ultimately, despite the widespread closure and/or the subdivision of sites, capacity utilisation declined from 22 per cent in 1978 to 17.6 per cent in 1983.[22]

It is tempting to think that the virtual collapse in cinemagoing was characterised by a constant decline; however, this was not the case. In 1954, there were still more than 1.27 billion admissions per year, signifying that the cinema remained an important form of entertainment. However, the fact remains that there were periodic and sudden drops, with the period after 1954 a case in point, as the audience had declined by more than half to 500 million by 1960. Thereafter, the rate of decline steepened and accelerated so that by 1970 admissions had more than halved again to 193 million and would halve again to 101 million by 1980. The people who visited the cinema most often were the young, specifically those between 7 and 24 years, and they were the group whose attendance remained fairly constant in the period of most dramatic decline in attendance from the 1970s onwards. In 1984, according to audience research, 73 per cent of 7- to 14-year-olds claimed to ever go to the cinema, whilst the figure for 15- to 24-year-olds was 59 per cent, as opposed to 38 per cent for all cinemagoers aged over 7.[23] It seemed as if the reliance on a young audience was based upon their attendance as habitual rather than out of a deep conviction to the cinema itself, especially given its parlous state. Ascanio Branca, managing director of 20th Century Fox (UK), set out the challenge:

> The young will go back to the cinema just to see a film, but the place of entertainment, which is the cinema, must be comfortable, must have atmosphere, and must have a good quality sound and projection, must be clean and decent, in fact must have all the features which will encourage not only the young to go out but couples and older people too … Quality pictures have now come back, so the cinema must now fulfil their part of the chain both in quality and quantity.[24]

What emerges from an examination of the decline in the numbers of cinemas is that the problems for the exhibition industry and the possible remedies were too often focused on the *mechanics* of exhibition and distribution, rather than on the social and cultural *experience* of film.[25] The exhibition industry missed what Docherty et al. identified as "one of the most important aspects of the decline in cinema attendance, namely the cultural shifts and changing material world of the audience upon which it depended."[26]

### LACK OF IMAGINATION AND INVESTMENT ON THE PART OF EXHIBITORS

As the exhibition industry shrank the two dominant companies—Rank and ABC (purchased by EMI in 1969, which became Thorn-EMI in 1979 and Thorn-EMI Screen Entertainment (TESE) in 1983)—effectively dominated British exhibition in the period up until the opening of Britain's first multiplex. A key issue, therefore, was accounting for the lack of investment in new cinemas by the Rank/Thorn-EMI duopoly and the extent to which this provided encouragement for new overseas multiplex owners, notably from the USA. From the 1960s onwards, this duopoly had introduced the concept of multiscreen cinema via the dividing and subdividing of its exiting sites. By 1984, Thorn-EMI's ABC circuit had shrunk to 107 sites and 287 screens, and the Rank circuit to 75 sites with 194 screens. Though at that point a smaller player, the Cannon Group, had with the purchase of the 67-site Classic cinema circuit in 1982 emerged as a not-insignificant third player, and which would become a major player with the takeover of Thorn-EMI's cinemas in 1986. In 1984, it was estimated that approximately a quarter of the population now lived 20 miles or more from a cinema.[27] Taking as the starting point Olins' analysis that as part of large conglomerates with interests in a variety of other fields, the duopoly had neglected their cinema businesses, the key seemed to be the enduring attraction of film and the demand for a different kind of cinema.[28] Commenting in 1985 on efforts to increase the cinema audience, the Association of Independent Producers observed that:

> the ground beneath its feet will be cut away unless the major exhibition circuits, owned by Thorn EMI, Rank and Cannon, make tangible commitments to improve the quality of their cinemas and of the sound and projection equipment in them. People simply will not pay to see films, however good, in dirty, smelly old cinemas. You'd have thought the message would have got through by now.[29]

Though Thorn-EMI did acknowledge that the cinema environment was important, they insisted that audience research, undertaken for them in 1983 by PA Consulting, indicated that the overwhelming reason given by 35 per cent of adults and 45 per cent of 15- to 19-year-olds for not visiting the cinema more often was that tickets were too expensive.[30] For its part, Rank had started the 1980s with a closure programme of 29 cinemas, citing a decline in admissions and disquiet amongst shareholders, though they sought to sweeten it by pledging to invest in updating and modernising the circuit's remaining sites.[31] In a paper entitled *Cinema Exhibition in the U.K.* distributed amongst members of the British Film and Television Producers Association, Bernard Kingham of ITC Entertainment observed that the economics of the exhibition industry were such that "the sort of massive rebuilding and refurbishment called for is difficult to sustain."[32] Olins' polemical study of Britain's cinemas offers a real sense of the state of the traditional exhibition sector on the eve of the multiplex and, having listed all of the supposed ills befalling the exhibition industry, including television, price, lack of good films, and video, Olins, whose company Wolff Olins had designed several cinemas, argued that what was wrong was simple: "[i]t's the cinemas. People don't want to go to cinemas because most cinemas are no longer pleasant places to be and they haven't been for a long time."[33]

As 1984 ended, admissions reached their nadir and, in an effort, to arrest the decline British Film Year was launched in 1985 with the twin aims to promote British films abroad and champion the cinema with the slogan "the best place to see a film." In a speech, Richard Attenborough contrasted a shrinking cinema infrastructure in Britain with that of the USA, where he claimed that 90 per cent of the cinemas had been built in the previous ten years, with a concomitant increase in admissions.[34] This attempt to raise the profile of British cinema and encourage people to see films in the cinema corresponded, ironically, with a spate of high-profile Hollywood studio blockbusters like *Beverley Hills Cop* (1984) and *Rambo: First Blood Part II* (1985). The growth in admissions of an unprecedented 31 per cent during 1985 appears impressive, but it marked a rise from the lowest point ever—only 3.3 per cent of the 1946 admission level.[35]

In his analysis of British Film Year, Higson offered a sense of the tension between a desire to reposition cinema as both a public medium, to be used to articulate and circulate versions of the nation, and the dominance of Hollywood.[36] Though not explicitly about multiplexes or any particular form of exhibition, Higson's article was important in determining the conditions of cinema and cinemagoing immediately prior to the multiplex's arrival. These conditions were explicitly concerned with the articulation of market capitalism under Margaret Thatcher's Conservative government, which saw a tendency towards Britain's developing role in a rapidly globalising economy. This, it is argued, is the critical moment for the development of a new form of cinema, based on the experiences of multinational US cinema companies. As Park observed: "[i]t's increasingly acknowledged that for the cinema, as for other retail sectors, the future lies in suburban complexes near to residential communities. About 50 free-standing multiplexes adjacent to key cities, it's argued, could turn the business around overnight."[37]

Prior to 1985, the pattern of cinema exhibition in Britain was characterised by contraction. In 1979, Laurie Marsh, President-elect of the Cinema Exhibitors Association and head of the Classic Cinema circuit, noted that "in the past 20 years, there have been virtually no new cinemas in positions where there were none before."[38] Given that in the period after 1985 the pattern was a steady increase in the diffusion of multiplex cinemas, a key issue, therefore, was accounting for the lack of investment in new cinemas by the Rank/ABC duopoly and the extent to which this provided encouragement for new overseas multiplex owners, notably from the USA.

## THE RANK/ABC DUOPOLY BECOMES THE RANK/CANNON DUOPOLY

Despite having a dominant market position, no British exhibition circuit was advocating the systematic and widespread building of new cinemas in the late 1970s and early 1980s; indeed ABC had not opened a new cinema since 1973. Moreover, as Eyles noted, in 1986 Rank chose to invest £67.5 million in the purchase of slot machine arcades and bingo halls.[39] As a company, Rank did not open its first Odeon multiplex—an eight-screen complex in Stoke-on-Trent—until October 1989. It was the 23rd multiplex opened in Britain. Rather than focus necessarily on larger 8- to 12-screen multiplexes, Rank professed an enthusiasm for "miniplexes" which were sites with up to five screens such as those opened in Ipswich and Cardiff in 1991 and Chelmsford in 1993.[40] In the case of the Odeon Leicester Square in London, Rank filled in a passageway on the side of the cinema and built five small auditoria of 50–60 seats, which opened as the Odeon Mezzanine in April 1990. The reluctance to invest in larger multiplexes in favour of upgrading and splitting sites meant, according to Dodona, that whilst the Odeon circuit's audience was not declining it was not increasing either.[41] Moreover, its "stringent investment criteria" and reluctance to leverage its multiplex developments meant that "the long term future of the circuit rests on how successfully and aggressively it is able to pursue its now clear strategy of seriously entering the multiplex market."[42] Rank's perceived reluctance to get into the multiplex market, which site development manager Dean Morton denied was the case, was in part determined by the siting of their majority of sites in town and city centres; well away from the multiplexes opening in out-of-town locations on the periphery.[43] Nonetheless, as Ilott observed, Rank "steered away from large-scale multiplex investment" because of the high costs involved, with Chief Executive Michael Gifford arguing that "building a multiplex is an expensive undertaking and the big multiplexes don't get the returns. Admissions yes, but returns no."[44] It was not until 1997–8 that Rank's multiplex openings exceeded those of its competitors.

ABC, founded in 1928, was part of Thorn-EMI and headed by an American, Gary Dartnall. Having seen AMC's proposed new multiplex in Milton Keynes, Dartnall had been keen on a multiplex building programme for his company. The site chosen for Thorn-EMI's first multiplex was Salford Quays in Manchester.[45] With an estimated cost of £3.5 million the eight-screen complex would have been ABC's first multiplex. Dartnall had to work hard to convince a sceptical Thorn-EMI Board for whom the location was the subject of some derision, being as it was a redeveloped former industrial area near the Manchester Ship Canal, and not a traditional city centre site.[46] In March 1985, Dartnall outlined his multiplex plans which then consisted of six cinemas, each with six screens, on sites adjoining shopping or leisure centres on the edge of cities and towns, with one executive finally admitting that converting existing cinemas was outmoded.[47]

Cannon had entered the British exhibition sector via its purchase of the Classic Cinemas circuit in 1982 from Lew Grade's Associated Communications Corporation, which had 67 sites and 130 screens. Cannon's owners Menahem Golan and Yoram Globus purchased the ailing Star cinema circuit's 37 sites in 1985 to make the newly named Cannon Cinemas the second biggest exhibitor in Britain.[48] Having broken the existing Rank/ABC duopoly, in a proposed partnership with Gerald Ronson's Heron International, Cannon set its sights on the acquisition of the newly designated Thorn-EMI Screen Entertainment's (TESE) ABC circuit, which had been put up for sale in 1985. Despite Dartnall's optimism about the future of the ABC circuit, 1984, as we have seen, had been the nadir of cinemagoing and ABC was particularly vulnerable. In May, George Lennox, director of operations at ABC, told *Variety* that "up to 35% of the present sites, many of them in blighted inner-city areas" were "no longer viable" and were "more valuable as real estate" than as cinemas.[49] There was much vocal resistance to Golan and Globus' attempts to acquire ABC,[50] which would have made Cannon the largest circuit in the country, not least from established figures within the British cinema establishment, including the Association of Cinematograph, Television and Allied Technicians (ACTT) and the Association of Independent Producers (AIP). For its part, the AIP focussed on the fact that TESE would fall into "foreign hands" and the perceived "autocratic" style of management exercised by Golan and Globus.[51] Most vocal, however, was producer David Puttnam who characterised himself as an "appalled onlooker" and described the potential sale of TESE to Cannon as "a disaster" for the domestic film industry, which would see Cannon dominating the market.[52]

The flak coming from Puttnam and others seemed to have worked as TESE's board saw Dartnall's proposed management buyout as the preferred option, especially as Rank were a bidder in addition to Cannon and Heron. Indeed, the board extended the deadline for purchase to allow

Dartnall to find a backer. In the event an Australian tycoon, Alan Bond, stepped in to support Dartnall's bid in December 1985 by providing the £10 million deposit Thorn-EMI wanted in order to approve Dartnall's buyout. However, unable to raise the necessary finance from his other backers, Dartnall's efforts to purchase the company failed and on 21 March 1986 the Bond Corporation acquired TESE for £125 million. One week later, Bond sold the company to Cannon for £175 million—a profit of £50 million—with Bond remaining a director of TESE as well as becoming a director of Cannon Group Inc.

Puttnam called for the intervention of the Monopolies and Mergers Commission (MMC) on the basis that Rank and Cannon were together responsible for 72.6 per cent of all film rental, describing himself as "always an implacable opponent of monopoly and duopoly."[53] In any event, since the new company would own more than 25 per cent of the market, the Office of Fair Trading (OFT) automatically investigated the acquisition and in July 1986 submitted its confidential report to the Department of Trade and Industry. In its report, the OFT recommended that the deal be referred to the MMC in the light of the increased market power of the Cannon group. However, the Secretary of State, Paul Channon, overruled the recommendation and approved the sale on the basis that the MMC would need to investigate the whole industry and not simply this sale.[54] Upon completion of the purchase, Cannon became the largest cinema exhibitor in Britain, with 200 cinemas and 485 screens.[55] In 1985, on the eve of the multiplex, Rank and Cannon controlled 54.3 per cent of screens, 64.1 per cent of admissions, and 66.5 per cent of revenues.[56]

## US Exhibitors Enter the British Market—Battling the "High-Street Dinosaurs"

As the multiplex was consolidated in North America, some of the major circuits—AMC, Cineplex-Odeon, and National Amusements—began to look to the European market for new opportunities. One of the markets identified for entry was Britain, which *Variety* described as "the most celebrated theatrical invalid among the major European territories."[57] Nevertheless, companies like AMC, Cineplex-Odeon, and National Amusements began to look to Britain for potential multiplex sites, based on the assessment that Britain was under-screened compared to other European countries and that the focus on city centres rather than the suburbs meant that "screens are mostly where the people are not."[58] National Amusements' Ira Korff summed up the optimism for expansion into the British market thus: "[w]e think that the reasons people go out to the movies in the US are the same things that would draw people in the UK."[59]

Amongst many of the initial wave of North American companies targeting Britain, the economic and political climate was both familiar to them and highly conducive to this new form of leisure-based industry. As Delmestri and Wezel argued the operation of multiplexes "required the integration of new knowledge, and the actors carrying this knowledge were traceable in the real estate or large-scale retail industry."[60] Moreover, the dominance of Britain's screens by Hollywood films reflected a broader climate of cultural connectedness with the USA. Britain's rapid adoption of the multiplex cinema, in contrast initially to most of Europe, lay in part to the extent to which the "cultural template of multiplexes" was less at odds with Britain's "cultural beliefs and practices."[61] Implicit in this "cultural template" was an acceptance that the multiplex's location within and in relation to the shopping or leisure complex was indicative of a changed cinema culture in which, Smith argued, there had been a "discursive meshing of their spatial conditions and consumption practices and that this has been a deliberate strategy of the multiplex industry … multiplex cinemas sell cinema itself."[62]

The introduction of multiplexes in Britain and other parts of the world by exhibitors which were closely tied to the major studios was part of a concerted effort to preserve the hegemony of Hollywood. With the arrival of multiplexes, British-based film distributors were persuaded to close up release patterns with the USA. As the number of multiplexes grew, this delay decreased further as distributors were offered more than one exhibition circuit for their releases.[63] However, the domination of the British exhibition sector by a duopoly of Rank and Cannon and a system of what was called 'barring'—denying other cinemas within a fixed distance of their cinemas getting first-run film prints—meant that initially the new multiplex companies like AMC were not always able to obtain prints of new releases.[64] Despite some efforts to lobby the British government, the European Commission, and the Monopolies and Mergers Commission (MMC), AMC and the other multiplex operators which entered the market in the first few years after 1985 demonstrated to distributors that the multiplex could not be ignored by the "High-Street dinosaurs."[65] By the early 1990s, alignments had been weakened as multiplexes entered the market.[66] Indeed, by the time the MMC examined the exhibition industry in 1994 the market penetration of multiplexes with their links to US-owned distributors was significant.[67] As the MMC observed:

> [a]ll the Hollywood studios rely upon their respective affiliates to distribute their films in the UK. Given the state of competition among the studios and in the distribution market generally, we do not object to this practice, which is common world-wide. Four of the seven Hollywood studios also have ownership links with UK exhibitors. We have examined whether these links result in dealings between distributors and exhibitors being other than at arm's length. Our analysis shows a slight degree of preference between vertically linked parties at the margin, but the evidence does not warrant an adverse public interest finding.[68]

Between 1985, when AMC opened The Point, and 1990, Cannon and Rank were well behind their US rivals in opening multiplexes, having fewer sites than just one US operator, United Cinemas International (UCI), and the same as Warner Bros.[69] In this first period of the multiplex's development the impetus lay almost entirely with North American exhibitors. As Dodona argued:

> The real role of the multiplex, of course, has never been solely to increase the audience but rather to make it possible for exhibitors to continue in business. Before the multiplex concept was introduced all exhibitors could, and did, do was try to spin out the lives of the cinemas they already had. This was a profitable business but not profitable enough to support new capital investment: indeed, the argument could be turned on its head to state that it was the lack of competitive pressure to make new capital investments that made it such a profitable business.[70]

## The Point—"A UK Cinema for the 80s and 90s"

In 1984, US parent company AMC Entertainment Inc. formed AMC International with the express intention of expanding into the overseas exhibition marketplace, in what Chief Executive Stanley Durwood identified as "the most aggressive way possible."[71] Countries identified included Canada, Australia, and several in Europe, and Britain in particular. AMC was the key driver of the multiplex revolution in Britain, with Durwood's company characterised as "exploding outward" by head of AMC

International, Robert Friedman.[72] In 1985, AMC made it clear that they were looking to invest "substantial sums and expected to see returns,"[73] a comment made whilst they were mid-way through the construction of Britain's first multiplex cinema in the Buckinghamshire new town of Milton Keynes.[74] Before settling on Britain AMC had looked at sites in Europe, especially Scandinavia. Answering the question "[b]ut why this development in Milton Keynes?" AMC's Charles Wesoky argued that the decline in cinemagoing in Britain was because cinemas were badly operated, were too expensive, and had inadequate travel facilities.[75] In July 1983, initial proposals were announced by a consortium of Bass Leisure Activities, a British brewing company, and AMC for an 80-foot crystal pyramid of mirrored, prismatic glass; a "glittering landmark for a 21st century entertainment centre in Central Milton Keynes."[76]

The partnership of Bass and AMC was in many ways a serendipitous one, as both were brought together by a desire on the part of Milton Keynes Development Corporation (MKDC) for a major entertainments complex. In 1982, MKDC identified an "entertainment package" to include facilities for cinemas, bingo, and a nightclub/discotheque[77] and set out to market a prime city centre site for a leisure and entertainment complex, which it defined as "bright lights' city centre entertainments."[78] Proposals would be invited from developers and operators for a bingo hall to cater for upwards of 1000 people, a 700-seat cinema, and a nightclub/ discotheque for some 800 people.[79] Bass felt strongly that a central attraction of their leisure complex should be a multiplex cinema; however, since no British-based exhibitor had any experience in this area they invited AMC to build it.

Rank, the major cinema exhibitor in Britain under its Odeon brand, did not tender for the new entertainments complex, since, according to the company's head Jim Whittell, there seemed no way in which the operation could make either a profit or adequate returns.[80] The new Bass/AMC venture was called MK Entertainments Company and it chose the name The Point with the motto "Where People Matter." Both companies were significant operators in their respective markets, with Bass Leisure Activities having a turnover of more than £570 million, whilst AMC was at that time the third largest cinema circuit in the USA with 800 screens in 156 multiplexes in 26 states of the USA.[81]

When The Point opened on 29 November 1985, it was Britain's first multiplex cinema and whilst it was not the first purpose-built multiplex in Europe, it was, significantly, the *first* American multiplex in a foreign country.[82] Construction had started in August 1984, at a cost of £7.7 million (of which AMC contributed some £2.5 million). The eventual design was a 70-foot, bright red, multilevel structure resembling a Mesopotamian Ziggurat, with an adjoining windowless silver block at the rear housing the multiscreen cinema, which had ten screens—two 156-seat auditoriums, two with 169 seats, four with 220 seats, and two with 248 seats. In addition, The Point had a 1500-seat bingo club, a 130-seat restaurant, a 140-seat brasserie, a club/discotheque that could operate as a live performance venue and which accommodated up to 400 people and what was described as an 'American-style bar.'[83] All told, the complex contained 70,000 square feet of floor space and could accommodate up to 5000 people at any time. In terms of day-to-day operation, AMC managed the cinema whilst Bass Leisure operated the rest of the complex. Speaking some 25 years after the opening of The Point, Mark Batey, chief executive of the Film Distributors' Association, observed that:

> Consumers had forgotten about cinema, really. There were the Bond films every couple of years, the early Spielberg adventure films, *Star Wars*, and not a great deal else. Strange to think about it now. There was no habit. A lot of traditional sites were hanging on by their fingertips.[84]

The incorporation of different forms of commercial leisure facilities was seen as integral to the complex's appeal. Floyd described The Point as a "one-stop entertainment centre,"[85] whilst AMC's Charles Wesoky heralded it as a "revolutionary step forward in cinema" as "well managed cinemas within integrated entertainment centres spells the future of British cinema."[86] For film producer David Puttnam, The Point was "a UK cinema for the 80s and 90s,"[87] whilst he described the opening as "one of the most exciting days of my life."[88] The incorporation of different forms of commercial leisure facilities was seen as integral to the complex's appeal. Bass Leisure's Peter Sherlock argued that "The Point can be the venue for a full day or evening out, customers don't need to decide what to do until they arrive."[89] Commenting on the development of The Point, John Wilkinson, Chief Executive of trade body the Cinema Exhibitor's Association, observed that "if you put the good projection, the good sound, the bigger seats, the popcorn, the post movie drinks and the better trained staff in one place instead of concentrating on just one of those elements the whole effect was greater than the parts."[90]

Of course, The Point was not Britain's first multiscreen cinema. For the previous 20 years, the major exhibitors had been subdividing their sites into two and three screens; however, this "twinning" and "tripling" of existing cinemas was seen as giving multiscreen cinemas a bad press. In contrast, the multiplex was seen as banishing this due to the design of the auditoria. AMC trumpeted The Point's ten screens as having "computerised sightlines," larger seats with more legroom, the latest Dolby stereo sound systems, and the widest possible screens.[91] The Point had three projection rooms (two for four screens and one for two screens) and in any one week could operate up to 300 film showings of some 14 different titles.[92] In common with subsequent multiplex developments, The Point installed automated projectors, manufactured by Italian company Cinemeccanica, which used large metal platters—metal discs some four to five feet in diameter stacked on top of each other in what is known as a "cake stand"—for the film rather than traditional reels. The separate reels of film that arrived from the distributor would be spliced together and wound onto the feed platter, from which the film would pass through the projector and onto the take-up platter. The advantages for the multiplex operators were that unlike traditional reels (approximately 2000 feet in length), which required changeovers and thus two projectors per screen, this system required only one projector per screen.

The capacity afforded by ten screens meant that AMC's main selling feature of the multiplex cinema was "choice" with two or more showings of individual films per evening and at least one screen showing a U certificate film for children.[93] AMC adopted a system of screenings in which the most popular new releases were timed to start first with others starting within 10–15 minutes, which meant that any cinemagoers unable to get into their first-choice film could find something else to watch. This staggered timing of films, characterised as "spill and fill," was supported by a dramatic new technological feature: the capacity to show one print on up to four screens, which AMC claimed would enable them to "keep abreast of viewing demands."[94] AMC stressed the importation of other successful elements of the American formula such as computerised ticket sales, credit card booking, new kinds of food and drinks, and a no-smoking policy throughout. Moreover, the design of the cinema meant easier access for disabled cinemagoers too, including the provision of suitable toilets. Pulleine reported that the cinema would "bear no relation to the insalubrious, hard-to-reach downtown cinema."[95]

AMC's vice-president for international theatre development, Scott Wallace, argued that The Point would re-establish cinema as a common-place form of entertainment rather than the "special event" that cinema had become in Britain.[96] Though professing to "play a hunch that had been wildly successful in the US," Wallace insisted that the development was not a risk since AMC had begun to research the British exhibition market in 1979, five years before they committed funds for the development of The Point.[97] Moreover, the negotiations with Bass Leisure Activities were underway for some two years before the deal to build The Point was announced. Part of the "hunch" involved looking for a site that was similar to those that had been successful in the USA, and in part this was the location—opposite the main shopping centre, fronting onto Midsummer Boulevard—and with a surrounding population who could find their way easily to the new complex and park. To this end, the new town of Milton Keynes was ideal. Amongst many innovative aspects of the plan of Milton Keynes was the layout of the city on a US-style grid system of roads, principally as an aid to what Clapson called "mobile and increasingly flexible living patterns."[98] The importance of the motorcar was cemented in 1979 with the opening of the US-style indoor shopping centre, which was subsequently extended in 1993. The centre proved a popular attraction not just for the inhabitants of Milton Keynes but also for those in surrounding towns and cities.

Bass Leisure's Peter Sherlock calculated that three quarters of the population was under 45 and affluent; moreover, The Point aimed to draw from a catchment area within a 15-mile radius inhabited by approximately 1.5 million people, all within a 45-minute car drive of the town centre.[99] Both Bass and AMC believed that such a complex would be very attractive to young people, as research suggested that 40 per cent of the population of Milton Keynes was under 25 years old.[100] Since the cinemagoing audience was dominated by under-25s, Milton Keynes seemed the perfect location. In the first year of operations, attendances exceeded one million and the cinema was generating some 21,000 admissions a week.[101] In an interview with Nick Glass on Channel 4 News (5 November 1986) Charles Wesoky stated that admissions in the first year of operation were 32 per cent above expectations. As Dodona pointed out:

> [b]y allowing the generation and realisation of significant economies of scale the multiplex raised the possibility of investing in cinemas again. This brought the major advantage of being able to relocate cinemas to take

advantage of shifts in population and changes in consumer habits, such as visiting shopping malls and even driving cars which had occurred since many cinemas had been built. It was no accident that the very first multiplex built in Europe was in a completely new city designed around the motor car.[102]

AMC set a target of building 30–40 multiplexes with screens totalling 300–400 by 1990, which if realised would have made AMC the second biggest exhibitor in Britain.[103] However, that was a wildly optimistic proposal and in reality AMC opened a further seven 10-screen multiplexes by 1989. In 1988, in order to finance their on-going operations, AMC formed a British-based but US-owned joint venture with two other US companies—Cinema International Corporation (CIC) jointly owned by Paramount and Universal, and United Artists Communications, which was already an active multiplex builder. The new company was called AMC (UK) Ltd. Each of the three companies had a 33 per cent stake in the new venture, investing £24 million. The plan was to raise a further £120 million from merchant banks in the city in order to re-structure debts and fund the opening of 37 new multiplexes by 1991.[104] However, within a few months, in December 1988, AMC announced that they were to pull out of the British market altogether, ostensibly they argued, to concentrate on their core North American business, which was believed to be experiencing financial difficulties.[105] They sold their existing multiplexes to CIC and United Artists Communications, which was renamed United Cinemas International (UCI) (see below). Three years after The Point opened the AMC brand disappeared in favour of UCI.[106] The two further multiplexes AMC were building in West Thurrock and Swansea were branded UCI when they eventually opened in 1989.

## US CIRCUITS ENTER A VOLATILE MARKET

Among the first major US cinema exhibitors to expand overseas, CIC was a vertically integrated company in Britain since Paramount and Universal also co-owned United International Pictures (UIP), a leading film distributor with some 18 per cent of the British market.[107] In 1986, CIC announced plans to build two multiplexes in High Wycombe and Solihull under their British trading company CIC Theatre Group, which already owned two large West End London cinemas—the Empire and Plaza. Additionally, CIC operated some 30 screens in South Africa in collaboration with MGM through its CIC-Metro circuit and had a development deal in Australia with exhibition circuit Hoyts.[108] In addition to Britain, CIC had also set their sights on expanding into the Scandinavian, South American, and European markets. In April 1987, CIC had opened Britain's third multiplex—the Wycombe 6—in High Wycombe, which seated 1728 and was at that point the most expensive yet built at £4.3 million.[109] CIC undertook localised research, including surveys of the local population, before building the Wycombe 6.[110] The cinema was located just off the M40 motorway and had a large population of approximately 700,000 within easy travelling distance. Adjacent to the multiplex was a large sports centre, a supermarket, and a large department store, all additional attractions for visitors. Though the proposed eight-screen Solihull site was announced at the same time as Wycombe, it was not opened until October 1989. In the same year, CIC was merged with American Multi-Cinema (AMC) to become United Cinemas International (UCI)—the largest multiplex cinema circuit in Britain at that time—and with the withdrawal of UAE, now joint-owned by Paramount Studios and Universal (see above). With the formation of UCI in 1989, CIC's sites were subsequently rebranded, whilst CIC acquired the whole of the UCI company when partners United Artists sold out their 50 per cent share in 1990. Like AMC before them, United Artists were seeking to stem significant losses in their US operations, which *Variety* characterised as "cash starved companies disposing of assets."[111]

Canadian multiplex pioneers Cineplex Odeon entered the British market in 1988 and like AMC, the company, which was one of the fastest growing circuits in North America, talked up their international ambitions boldly. The company's British subsidiary, Cineplex Odeon (UK), planned to invest over £50 million in building 100 screens with financial support from Crédit Lyonais, whilst its flagship proposal was a four-screen cinema in London's recently renovated Trocadero Centre. Unlike AMC and CIC, Cineplex Odeon professed an intention to focus on the area around London, within the M25 catchment and build smaller six- to ten-screen complexes.[112] As Chief Executive Garth Drabinsky revealed, "the London area is where our primary focus will be," adding that "I don't have a huge intention of building theatres in areas with 150,000 people 100 miles outside of London. We want to attack the main markets in the UK first of all."[113] Cineplex Odeon's first multiplex, though, was the result of the takeover of an existing company—Maybox Group plc.—and the acquisition of its ten-screen complex in Slough, in June 1988. Maybox had been the second largest operator of live theatres in London, and the Slough multiplex, opened in September 1987, had been the first of a planned series of five to ten multiplexes over the following five years. In the event the company sold out to Cineplex Odeon (UK) as they considered themselves "too small to compete in the marketplace."[114] Press reports though seemed to suggest that the multiplex in Slough had "underperformed" in comparison to others and the Wycombe 6 in particular. Reasons cited included the location and lack of car parking.[115]

Cineplex Odeon (UK) was forced to rebrand itself as Gallery Cinemas in 1989, in part because the Odeon name had been used by Rank for its circuit for many years. Whilst it did not own Gallery outright, it controlled 55 per cent of the company with a variety of British investors owning the balance. At this point, it had only one multiplex in Slough but set out ambitious plans for a series of multiplexes to be opened in 1990, including six-screen cinemas in Brighton, Brent Cross, Eastbourne, Harlow, Glasgow, and Gloucester, in addition to the cinema then under construction at the Trocadero in London. As well as their London site, the first to be started was in Brighton, whilst both were partnerships with the developer Brent Walker, which owned Elstree Studios and film producer Goldcrest. Like these, many of Gallery's proposed cinemas were integrated into larger leisure developments, such as Eastbourne's Crumble Centre, the company argued, mainly because of the "synergies" it afforded "particularly shared car parking."[116]

Before Gallery cinemas could open any of its cinemas parent company Cineplex-Odeon sold the fledgling circuit to the Cannon Group, the largest exhibitor at that time, for approximately £20 million in March 1990.[117] Like AMC, Cineplex-Odeon's decision was largely dictated by its business in North America, where it had made a loss of $78.6 million in 1989.[118] In the wake of a major corporate restructuring the company, itself owned by media giant MCA Inc., had sought to dispose of its overseas assets in order to reduce its debts. Cineplex Odeon was also unable to finance its proposed building of 100+ screens in 45 locations since, although it raised £45 million, the increased costs of land and construction stymied its ambitions.[119] In the end, the company never opened a multiplex under its Gallery brand.

Unlike companies such as AMC and Cineplex Odeon, Boston-based National Amusements adopted a more careful and measured expansion into the British exhibition market. In 1987, National Amusements (UK) Ltd began construction of its first multiplex in Nottingham under its Showcase brand, the first of seven multiplexes built in the first two years of the company's entry into the British market. Shari Redstone, Executive Vice President of National Amusements, reasoned that Britain was "under-screened" and that the company could be "aggressive" in its expansion.[120] In an interview, Ira Korff, managing director of National Amusement (UK) Ltd, set out the rationale for the company's entry into Britain and the attraction of the multiplex for British cinemagoers:

> I decided to expand into the UK based upon the belief that the sort of lux-ury entertainment complex that we would build would be appreciated and enjoyed by the UK public and would draw them back to the movies. That it is, in and of itself and the film aside, an exciting experience to walk into these thousands of square feet of foyer space and then have any one of 11 first run films to choose from.[121]

National Amusements departed from the prevailing orthodoxy in a number of ways in their building. Unlike AMC and Cineplex-Odeon, they purchased their properties freehold rather than leasing and, unlike all three of the other US exhibitors discussed thus far, they opted exclusively for stand-alone complexes, which were not opened in conjunction with developers of associated entertainments such as shopping centres or lei-sure parks. Most of the multiplexes built in Britain in the first ten years were leasehold tenants within a larger development, which meant that they were particularly sensitive to increases in rental charges.[122] To this end, National Amusements tended to site their cinemas adjacent to main roads and intersections on the edge of major conurbations, and opted for fairly standardised "modular" designs, which allowed for the addition of two or three auditoria if demand increased sufficiently. Though National Amusements opened their first multiplex only three years after AMC's Milton Keynes complex, unlike AMC they have consistently maintained their presence in the exhibition market right through until the present day. By 1990, they had opened seven multiplexes, ranging from 12 to 14 screens and with seating for 2600–3500 people.[123] After ten years of oper-ation, the company had 15 sites with a total of 197 screens[124] and by 2017 they had 21 sites with a total of 278 screens, which was 6.5 per cent of total screens.[125]

One of the characteristics of the first five years of the multiplex's diffu-sion in Britain was the apparent fragility of several companies' position in the market, as witnessed by the rapid changes in ownership and number of takeovers and acquisitions. In 1986, the second year of multiplex developments in Britain, there were three companies operating purpose-built multiplex cinemas in Britain: AMC, CIC, and Cannon. They were joined by Showcase (National Amusements) and Cineplex Odeon (Gallery) in 1988. However, by 1990 only the Showcase brand remained; the others having been taken over or merged with new brands, whilst Rank did not open its first purpose-built Odeon multiplex until 1989. Nonetheless, having paved the way, the retreat of the first wave of US exhibitors was followed by the entry of some more US companies, such as Warner Bros. in 1989, but which also allowed for the emergence of a group of domestic circuits that would come to dominate the market by the 2000s. The real revolution though was the hegemony of the purpose-built, multiscreen cinema that was first the multiplex and subsequently the megaplex.

## Notes

1. Nick Roddick, "If the United States spoke Spanish, We Would Have a Film Industry…," in *British Cinema Now*, eds. Martyn Auty and Nick Roddick (London: British Film Institute, 1985), 8.
2. "British Cinema Fast Disappearing," *Screen Digest* (October, 1984), 187.
3. "British Cinema and Film Statistics," *Screen Digest* (October, 1990), 1.
4. David Docherty, David Morrison, and Michael Tracey, *The Last Picture Show: Britain's Changing Film Audience* (London: British Film Institute, 1987).
5. Cmnd 6372. *The Future of the British Film Industry: Report of the Prime Minister's Working Party*, The Terry Report (London: HMSO, 1976).
6. See Sian Barber, "Government Aid and Film Legislation: 'An Elastoplast to Stop a Haemorrhage'," in *British Film Culture in the 1970s: The Boundaries of Pleasure*, eds. Sue Harper and Justin Smith (Edinburgh: Edinburgh University Press, 2012) and Margaret Dickinson and Sarah Street, *Cinema and State: The Film Industry and the Government 1927–1984* (London: British Film Institute, 1985).
7. Sue Summers, "'Save UK Cinema' Plea to State," *Screen International* 98 (30 June, 1977), 1.
8. Ibid., 1.
9. "British Admissions Slump; 'Market' by Gut Instinct; A Nation of Home-Owners," *Variety* 301, no. 11 (14 January, 1981), 321.
10. Ibid., 170.
11. David Hovatter, "Where the Cinema Is Going Wrong," *Screen International* 339 (17 April, 1982), 19.

12. Tina McFarling, "Survey Points to Customer Dissatisfaction in UK Cinemas," *Screen International* 511 (24 August, 1985), 12.
13. Ibid., 12.
14. John Hogarth, "Building Admissions above the 50 million Mark," *Screen International* 483 (9 February, 1985), 25.
15. "British Cinema Fast Disappearing," 188.
16. Ibid., 187.
17. Simon Blanchard, "Cinema-Going, Going, Gone?," *Screen* 24, nos. 4–5 (July-October, 1983): 109–113, 109.
18. "BFI" compiled by Linda Wood. *The British Film Industry*, Information Guide No. 1. October (London: British Film Institute, 1980), 2.
19. "British Cinema Stats," *Variety* 329, no. 13 (20 January, 1988), 122.
20. Chris Brown, "British Cinemas Booming, But Can They Keep the Customers Happy?," *Screen International* 159 (7 October, 1978), 7.
21. David Sexton, "The Empire Strikes Back," *Sunday Telegraph Magazine*, 12 March, 1989.
22. "British Cinema Fast Disappearing," 190.
23. "British Cinema and Film Statistics," 8.
24. Brown, "British Cinemas Booming," 7.
25. Stuart Hanson, *From Silent Screen to Multi-Screen: A History of Cinema Exhibition in Britain Since* 1896 (Manchester: Manchester University Press, 2007).
26. David Docherty, David Morrison, and Michael Tracey, "Who Goes to the Cinema?," *Sight and Sound* 55, no. 2 (Spring, 1986): 81–85, 82.
27. "Brit. Admissions Up 10% In '83, But Decline Plagues Industry," *Variety* 315, no. 2 (9 May, 1984), 471.
28. Wally Olins, "The Best Place to See a Film?," *Sight and Sound* 54, no. 4 (Autumn, 1985): 241–4.
29. "News—British Film Year," *AIP & Co.* 59 (October, 1984), 15.
30. "Reformation of UK Film Exhibition," *Screen Digest* (February, 1984), 28.
31. Adrian Hodges, "650 Jobs Will Go as Rank Closes Cinemas," *Screen International* 298 (27 June, 1981), 10.
32. "ITC's Kingham Cites Ills, Possible Aid for Theatres," *Variety* 308, no. 12 (20 October, 1982), 266.
33. Olins, "The Best Place to See a Film?," 241–2.
34. Richard Attenborough, "Momentum of British Film Year Can Carry Effects Beyond 1986," *Screen International* 492 (13 April, 1985), 12.
35. "Major Cinemagoing Countries Collated," *Screen Digest* (October, 1986), 207.
36. Andrew Higson, "The Discourses of British Film Year," *Screen* 27, no. 1 (January–February, 1986), 86–109.

37. James Park, "U.K. Circuit Expansion, Renovation Thought to be Cure for Ailing Biz," *Variety* 316, no. 10 (3 October, 1984), 49.
38. Laurie Marsh, "The British Exhib's View of the Cinema's Decline," *Variety* 293, no. 9 (3 January, 1979), 98.
39. Allen Eyles, *Odeon Cinemas 2: From J. Arthur Rank to the Multiplex* (London: British Film Institute, 2005), 149.
40. See Laurie Clarke, "Odeon Cinemas," *Screen International* 891 (22 January, 1993), 19.
41. Dodona, *Cinemagoing 4* (Leicester: Dodona Research, 1995).
42. Ibid., 29.
43. See Allen Eyles,. "Odeon at 65," *Variety* 359, no. 2 (8 May, 1995), 46–48, 56 and Eyles, *Odeon Cinemas 2*.
44. Terry Ilott, "Rank Motto: Know When to Say No," *Variety* 343, no. 4 (6 May, 1991), 356.
45. By the time the multiplex at Salford Quays opened in December 1986 it was, however, a Cannon. The ABC brand had been consigned to history.
46. See Allen Eyles, *ABC The First Name in Entertainment* (London: British Film Institute, 1993).
47. Cited in Nigel Floyd, "View to a Screen Killing," *Stills* (19 May, 1985), 23.
48. See Philip Turner, *Cannon Cinemas: An Outline History* (St. Paul's Cray: Brantwood Books, 1997).
49. Jack Pitman, "About One-Third of ABC Cinemas No Longer Viable, Asserts Lennox," *Variety* 219, no. 2 (8 May, 1985), 92.
50. See Geoffrey Macnab, *Delivering Dreams: A Century of British Film Distribution* (London: I.B. Tauris, 2016).
51. Roger Watkins, "Cannon Grabs Screen Entertainment," *Variety* 323, no. 2 (7 May, 1986), 517.
52. Andrew Cornelius, "Film Industry Rallies Against Sale," *The Guardian*, 30 November, 1985.
53. *Variety*, "Puttnam Opposed to Cannon Buy of ABC Chain as Duopoly Reprise," 323, no. 4 (21 May, 1986), 5.
54. See Andrew Cornelius, "DTI Clears Cannon Bid Against OFT Advice," *The Guardian*, 14 August, 1986.
55. Philip Turner, *Cannon Cinemas*, 11.
56. "British Cinema and Film Statistics," 2.
57. "Exhibition on Upswing Globally," *Variety* 320, no. 6 (4 September, 1985), 90.
58. "What to Do about U.K. Cinemas, Now the Worst off in Europe," *Variety* 317, no. 12 (16 January, 1985), 100.

59. Quoted in John Hazelton, "Korff Breaks New Ground," *Screen International* 658 (25 June, 1988), 24.
60. Giuseppe Delmestri and Filippo Wezel, "Breaking the Wave: The Contested Legitimation of an Alien Organizational Form," *Journal of International Business Studies* 42 (2011): 828–52, 831.
61. Ibid., 831.
62. Justin Smith, "Cinema for Sale: The Impact of the Multiplex on Cinema Going in Britain, 1985–2000," *Journal of British Cinema and Television* 2, no. 2 (2005): 242–55, 246.
63. See Hanson, *From Silent Screen to Multi-Screen*.
64. See Adam Dawtrey, "The Resurrection of British Cinema," *The Hollywood Reporter International Exhibition Special Report* 317 no. 25 (May, 1991), S-14, S-19.
65. Chris Brackhurst, "Multi-Screens in Battle with High-Street 'Dinosaurs'," *Sunday Times*, 6 August, 1989.
66. See Monopolies and Mergers Commission, *Films: A Report on the Supply of Films for Exhibition in Cinemas in the UK*, Cm 2673 (1994).
67. See Ralf Ludemann, "UK Preview Exhibition: Made in America," *Screen International* 1940 (14 January, 1994), 17, 24.
68. Monopolies and Mergers Commission, *Films: A Report on the Supply of Films for Exhibition in Cinemas in the UK*, 1.13, 5.
69. Dodona, *Cinemagoing* (Leicester: Dodona Research, 1989), 21.
70. Dodona, *Cinemagoing Western Europe* (Leicester: Dodona Research, 2000), 6–7.
71. "AMC Broadens Commitment to 'Aggressive' Expansion," *Screen International* 475 (8 December, 1984), 6.
72. "AMC Forms Global Arm for Overseas Expansion," *Variety* 317, no. 6 (5 December, 1984), 7.
73. Alex Sutherland, "AMC Unveils Cinema Investments Plan as UK Admissions Show Big Gains," *Screen International* 492 (13 April, 1985), 2.
74. See Stuart Hanson, "A 'Glittering Landmark for a 21st Century Entertainment Centre': The Story of The Point Multiplex Cinema in Milton Keynes," *Historical Journal of Film, Radio and Television* 33, no. 2 (2013): 270–88.
75. Ian McAsh, "Take 5: People in Camera," *Films on Screen and Video* 5, no. 6 (June, 1985), 14–15.
76. MKDC, *Pyramid landmark for Milton Keynes entertainment*: Press release (20 July, 1983). Milton Keynes Development Corporation, Information Unit. The initial design proved too costly and was subsequently altered. See Hanson, "A 'Glittering Landmark for a 21st Century Entertainment Centre'."

77. MKDC, *A Recreation Strategy for Milton Keynes*, 8 February. Milton Keynes Development Corporation, EMC 201 (1982), 12.
78. MKDC, *Milton Keynes City Centre Entertainment: A Development Activity* (Milton Keynes Development Corporation, 1982), 14.
79. Ibid., 14.
80. Eyles, *Odeon Cinemas 2*, 157.
81. "AMC Forms Global Arm For Overseas Expansion," *Variety* 317, no. 6 (5 December, 1984), 7.
82. See Hanson, "A 'Glittering Landmark for a 21st Century Entertainment Centre'."
83. Bass Leisure/AMC, *2,000 Seats, 10 Auditoria and Unique Programme Scheduling—The AMC Multiplex Opens in Milton Keynes*: Press Release (21 November, 1985).
84. Quoted in Phil Hoad, "How Multiplex Cinemas Saved the British Film Industry 25 Years Ago," *The Guardian* 11 November, 2010.
85. Floyd, "View to a Screen Killing," 23.
86. Bass Leisure/AMC, *Milton Keynes 'Points' to the Future*: Press Release (21 November, 1985).
87. David Puttnam, quoted in "AMC Target is 40 Multiplexes by 1990," *Screen International* 525 (30 November, 1985), 2.
88. "Puttnam's applause," *The Citizen*, 28 November, 1985.
89. Bass Leisure/AMC, "Milton Keynes 'Points' to the Future."
90. Quoted in Ian Freer, "Uniplex!" *Empire* 99 (September, 1997), 101.
91. Bass Leisure/AMC, "2,000 Seats, 10 Auditoria and Unique Programme Scheduling."
92. Martin Fodor, *AMC the Point 10 Cinemas: Notes from the First AMC 'Customer Focus Group'* (Milton Keynes Urban Studies Centre, 1987).
93. The programme for the first week of operation (opening Friday 29 November, 1985) was: *Santa Claus* (certificate U), *Goonies* (PG), *Prizzi's Honour* (15), *St Elmo's Fire* (PG), *The Bride* (15), *Weird Science* (15), *My Beautiful Laundrette* (15) and *Back to the Future* (from 4 Dec) (PG).
94. Bass Leisure/AMC, "2,000 Seats, 10 Auditoria and Unique Programme Scheduling."
95. Tim Pulleine, *Sight and Sound* 54, no. 3 (Summer, 1985), 154.
96. John Cunningham, "Crystal Hall," *The Guardian*, 21 July, 1983.
97. Ibid.
98. Mark Clapson, *A Social History of Milton Keynes* (London: Frank Cass, 2004), 43.
99. McAsh, "Take 5."
100. Ibid., 14.
101. Peter Freedman, "Supermarket Cinema Cashes in," *The Observer*, 8 November, 1987.

102. Dodona, *Cinemagoing Western Europe* (Leicester: Dodona Research, 2000), 6–7.

103. "AMC target is 40 multiplexes by 1990," *Screen International* 525 (30 November, 1985), 1.

104. John Daniel, "US Challenge to Cinema Chains," *The Guardian*, 14 October, 1988.

105. Dodona, *Cinemagoing* (Leicester: Dodona Research, 1989).

106. In 2001 AMC re-entered the British multiplex cinema market, opening two new multiplexes in mixed-use developments at Birmingham's Broadway Plaza (a converted children's hospital and adjoining site) and in Manchester's Great Northern development (a converted railway goods warehouse).

107. "British Cinema and Film Statistics," 14.

108. See Arnold Shepperson and Keyan Tomaselli, South Africa. In *The International Movie Industry*, ed. Gorham Kindem (Carbondale: Southern Illinois University Press, 2000).

109. Tina McFarling, "Wycombe 6," *Screen International* 610 (25 July, 1987), 16.

110. See Hanson, *From Silent Screen to Multi-Screen.*

111. Don Groves, "Yank Cinema Titan UA Selling Blighty Interest to Partner CIC," *Variety* 339, no. 1 (11 April, 1990), 4.

112. Philip Turner, *Cineplex Odeon: An Outline History* (St. Paul's Cray: Brantwood Books, 1998).

113. Brown Chris, "West End Linchpin for Cineplex UK Subsidiary," *Screen International* 661 (16 July, 1988), 2.

114. "Britain's Maybox Group to Exit Exhibition: Competish Too Tough," *Variety* 331, no. 6 (1 June, 1988), 6.

115. Turner, *Cineplex Odeon.*

116. "Cineplex Odeon's British Subsid Gallery Hopes to Have 110 Screens or More by '91," *Variety* 335, no. 3 (10 May, 1989), 178.

117. Don Groves, "Cineplex Odeon Sells Its Gallery Circuit to Pathé Cannon," *Variety* 338, no. 11 (28 March, 1990), 2.

118. "Cineplex Sells Off 'Non-core' Business," *Screen International* 750 (31 March, 1990), 4.

119. Dodona, *Cinemagoing.*

120. Beth Porter, "New England Comes to England: National Amusements Continues Its United Kingdom Expansion," *Film Journal International* 100, no. 1 (1 February, 1997), 14.

121. Hazelton, "Korff Breaks New Ground," 24.

122. Sumner Redstone, "10 Years of Showcase Cinemas: Building Blocks—Promotional Feature," *Screen International* 1162 (12 June, 1998), 30.

123. "United Kingdom Cinema Market," *Screen Digest* (July, 1991), 160.

124. "Screen Finance Analysis of UK Multiplex Developments in 1997," *Screen Finance*, 5 February, 1998, 8.
125. BFI, *Distribution and Exhibition*. BFI Research and Statistics, August, 2018. https://www.bfi.org.uk/sites/bfi.org.uk/files/downloads/bfi-distribution-and-exhibition-2018-10-03.pdf (accessed 27 November, 2018).

# CHAPTER 6

# The Multiplex is Consolidated in Britain

Between 1985 and 2018, there has been a year-on-year increase in the number of both cinema screens and multiplexes, though the total number of cinemas has fluctuated. The development of the multiplex took place in the context of a series of waves and even building "booms," when the number of openings averaged more than one a month. Given the decline of cinemas in the period prior to the opening of The Point in 1985, the question is *why* at this point was the first multiplex cinema built and whether the downward spiral of cinema attendance was reversed *as a result of* the introduction of the multiplex cinema. Significantly, the subsequent development of the multiplex has run in tandem with a year-on-year increase in cinema admissions.

The first period of accelerated growth in multiplexes, in the period from 1985 to 1995, was characterised by a range of common features: many were built by vertically integrated film companies that were primarily US-owned; many were closely associated with the development of out-of-town leisure and shopping complexes; their construction was encouraged by a range of financial incentives that were associated with the regeneration of post-industrial Britain; many were built by cinema circuits who were subject to a succession of takeovers and mergers; and many were functional buildings with little architectural merit and which were not designed to endure.[1]

In 1990, the exhibition market in Britain was dominated by the duopoly of Cannon, with 27.6 per cent of screens, and Rank (Odeon), with

© The Author(s) 2019
S. Hanson, *Screening the World*,
https://doi.org/10.1007/978-3-030-18995-2_6

15.7 per cent of screens; with new player United Cinemas International (UCI) in the third place with 10.7 per cent of screens.[2] Ten years later in 2000, the largest exhibitors were Odeon, though no longer owned by Rank, with 21.2 per cent of screens, French company Union Générale Cinématographique (UGC) with 11.8 per cent of screens, UCI with 11.6 per cent of screens, and Warner Bros. with 10.8 per cent of screens.[3] The changes in ownership over this ten-year period reflected a period of change, with new circuits emerging, such as MGM and Virgin, and then disappearing as they were acquired by competitors. However, this period of change does not compare to the considerable volatility and concentration of ownership that took place in the first half of the 2000s, when by 2005 three companies—Odeon/UCI, Cineworld, and Vue—controlled 60 per cent of screens and operated 174 multiplexes.[4] These three companies then dominated the market right through to 2018, though with Cineworld assuming the first place ahead of Odeon and Vue.

## THE BEGINNINGS OF DOMESTICALLY OWNED MULTIPLEX CIRCUITS

As one half of the duopoly that dominated exhibition prior to the multiplex, Rank did not follow the lead set by companies like American Multi-Cinema (AMC) in building multiplexes initially. When it opened its first in Stoke-on-Trent in 1989 it was part of an expansion plan announced in May 1987 to add 200 screens in 75 locations, which included proposed multiplexes in Romford in Essex and a former AMC scheme in Hull, but was largely concerned with adding screens to existing sites in towns and cities such as Ayr, Brighton, Harlow, and Sheffield.[5] Its focus upon its existing traditional sites meant that whilst Rank would consistently lay claim to being the most profitable exhibitor it was, as Dodona, argued, largely the result of the reluctance to invest in new cinemas.[6] Though in 1993 the company operated 76 sites, only ten were multiplexes, with 67 out of a total of 326 screens.[7]

Odeon's major rival emerged as United Cinemas International (UCI), which, as we have seen (see Chap. 5), was formed out of a merger between Cinema International Corporation (CIC) and AMC in December 1988, and was joint-owned by MCA Inc./Universal and Paramount Communications.[8] It began rebranding the nine CIC and AMC multiplexes it inherited, including The Point in Milton Keynes, almost immediately and instigated a short but rapid expansion, so that between 1988 and 1990, it opened ten multiplexes with a total of 95 screens.[9] At a time of intense competition in the first half of the 1990s, UCI was the most prolific multiplex builder in the British market and by 2000 was operating 33 multiplexes and 338 screens.[10] Though its sites were almost exclusively in out-of-town shopping centres, such as The Metro Centre in Gateshead and Lakeside in Kent, UCI opened two multiplexes—Whiteleys in London's Bayswater and The Printworks in Manchester (see Chap. 7)—that were significant because of their locations in the centre of cities. The eight-screen Whiteleys, opened in December 1989, was part of the development which occupied the former Whiteleys department store. This followed a more recent trend in the USA for cinemas to be located in prestigious downtown shopping complexes, such as Chicago's Water Tower. UCI talked up the importance of Whiteleys as an example of the ways in which multiplexes could reinvigorate the London cinema landscape and add screens, predicting that it would draw some 780,000 patrons from a population of 1.5 million residing within a ten-minute commute.[11] However, as UCI were in the early stages of building their new multiplex, rival exhibitor Cannon announced that its two cinemas in the vicinity—the three-screen Cannon Bayswater and the four-screen Cannon Edgware Road—were to close and be sold for redevelopment, in July 1988.[12]

Having expanded quickly in the first half of the 1990s, UCI was much more cautious in the second half, opening multiplexes at a slower rate, so that apart from its Picketts Lock site in London's Lee Valley, opened in March 1994, it did not build any multiplexes between 1992 and the end of 1997. In part this was because of efforts to change the design ethos of their sites, but mainly because of the dramatic expansion of the company overseas as part of a five-year plan. UCI had expanded into mainland Europe early on and by 1995 had multiplexes in Austria, Spain, Ireland, and Germany. In 1990, UCI had opened Germany's first purpose-built multiplex in Cologne (see Chap. 9) and by the end of the decade, UCI had opened a further 19 multiplexes in Germany with 175 screens. UCI also targeted Japan with plans in 1995 to invest $400–500 million in 20–25 multiplexes within ten years and opening its first, the seven-screen Otsu site in Kyoto, in November 1996 (see Chap. 11).[13] By way of example of UCI's expansion, in October 1997, the company opened four multiplexes with a total of 35 screens in Berlin, (Germany), Graz (Austria), Kanazawa (Japan), and Cardiff (Britain).

In September 2004, Odeon Cinemas was purchased by private equity group Terra Firma Capital Partners Ltd, followed by UCI just a month later. Odeon Cinemas had been traded twice since 2000, when it was purchased from Rank by private equity group Cinven, which undertook a rationalisation of the new combined circuit, opening a series of multiplexes and closing some traditional sites, and then in 2003 when it was sold to an investor consortium led by German bank WestLB. Upon purchase of the two exhibitors, Terra Firma retained the Odeon name for its cinemas in Britain and re-branded all UCI multiplexes accordingly, whilst in Europe the UCI brand was retained. The Office of Fair Trading (OFT) considered the merger between Odeon and UCI, especially as the new company controlled 25–35 per cent of cinema exhibition market—as measured by the box-office, number of screens, or number of seats—and 17 per cent of sites in Britain.[14] It concluded that whilst there were no issues of loss of competition at a national level, there were issues with a loss of competition at a local level, specifically in 11 localities covering 13 UCI multiplexes where the new company's cinemas overlapped.[15] In these circumstances, the new company would have to divest itself of 11 sites in cities and towns such as Glasgow, High Wycombe, Poole, and Newcastle/Gateshead.

Odeon/UCI assumed the position as the largest exhibitor in Britain with a combined total of 130 sites and 936 screens in 2004 and for the next 12 years the company maintained its position as the leading exhibitor in terms of sites and screens.[16] In July 2016, the company was sold by Terra Firma to AMC which was itself owned by the Dalian Wanda Group, for £921 million, in a deal prompted in part by the sharp drop in the pound in relation to the dollar after the Brexit vote. Moreover, the acquisition of Odeon/UCI meant that the Dalian Wanda Group became the largest cinema owner in the world (see Chaps. 4 and 11). Adam Aaron, chief executive of AMC, described the acquisition of Odeon/UCI as a "once-in-a-generation opportunity to acquire Europe's leading cinema circuit and create the world's biggest and best theatre operator."[17] Within a few months of purchasing the circuit, AMC changed the name to ODEON Cinema Group.

As one half of the exhibition duopoly, Cannon had dominated the industry in the first half of the 1980s; however, by the start of the 1990s the rapid expansion, based on large-scale acquisitions (see Chap. 5), meant that the company experienced considerable cash flow problems. As part of a vertically integrated company, the cinema circuit was susceptible to the ups and downs of Golan-Globus' production arm and in 1986, the US Securities and Exchange Commission (SEC) launched an enquiry into Cannon's accounting of profits.[18] In the aftermath, the company was effectively rescued by a $200 million loan organised by the flamboyant Italian financier Giancarlo Parretti, who in 1987 had begun purchasing various arms of the Cannon empire including the cinemas and Elstree studios.[19] By 1990, Parretti, head of Pathé Communications Corporation, which now controlled Cannon cinemas, orchestrated the purchase of MGM studios for $1.3 billion with finance provided by French investment bank Crédit Lyonnais' Dutch subsidiary CLBN.[20] MGM International Cinemas Group was formed with circuits in the Netherlands and Denmark amongst other European countries (see Chap. 8), with MGM Cinemas (UK) being the domestic arm. Slowly, the Cannon brand was replaced with the MGM logo, though this tended to be the multiplexes, whilst all new builds, such as the nine-screen Arcadian Centre in Birmingham, built in 1991, were opened as MGMs.

By 1994, MGM Cinemas (UK) Ltd was Britain's largest exhibition company and the second largest multiplex operator, controlling approximately 25 per cent of the market share.[21] In January 1993, Managing Director Barry Jenkins wrote, alluding to MGM's logo of Leo the Lion, that "it is an undisputed fact that the lion is the king of the jungle. It is also a fact that Leo is the king of the exhibition industry."[22] As one of the dominant exhibitors and distributors in the British market, MGM came to the attention of regulators. In 1994, the Monopolies and Mergers Commission (MMC) looked again at the exhibition industry and found against all of the leading exhibition companies, arguing that their relationship with the major distributors was a restriction of trade.[23] Moreover, the MMC found that legally there existed a complex monopoly in exhibition with MGM and Odeon, both of whom were vertically linked exhibitors, having more than 25 per cent of the market (57.2 per cent in actual fact) and that both companies were engaged in practices which limited competition. Nevertheless, the MMC concluded that despite some issues with alignments (offering films to particular cinema circuits with which the distributor has an agreement) and minimum exhibition periods for films: "[t]he transformation of the industry over the last decade has worked well for consumers … Today, apart from the practices we have criticized, competition in the markets under investigation is effective."[24]

Nevertheless, despite the clean bill of health, in 1992, Parretti was ousted having got into considerable financial difficulties and defaulting on loan repayments, which meant that Crédit Lyonnais acquired the newly renamed MGM Inc. and its international cinema exhibition circuit, including MGM Cinemas (UK) Ltd. Later that year the studios were sold off with Crédit Lyonnais retaining ownership of MGM International Cinemas Group. The costs of maintaining and upgrading the circuit's more traditional one-, two-, and three-screened Cannon cinemas in Britain's towns and cities were considerable, and the emphasis was placed on the more profitable sites. Nonetheless, Crédit Lyonnais invested in Britain's MGM Cinemas, opening new multiplexes in Belfast and Northampton, and planning new sites in Rochester, Leeds, and Aberdeen, amongst others. However, Crédit Lyonnais were struggling financially, having made a loss of FRF6.9 billion in 1993, and, despite repeated assurances that the cinema circuit was not for sale, the company was finally put on the market in October 1994.

There ensued a range of offers for MGM Cinemas, including one from Rank, Britain's largest exhibitor, and a proposed management buyout led by MGM's managing director Mike Sommers. In the event the successful buyer was Richard Branson's Virgin Group, which in a consortium that included private equity company Texas Pacific Group (TPG) and investment company Colony Capital paid approximately £195 million in July 1995, though Virgin's share was only £20 million. Virgin was a group of companies largely focussed on retailing and air travel and the purchase of MGM was a re-entry into the film business since Virgin Films had been a major producer in the early 1980s.[25] Like so many of the takeovers in this period, Virgin Cinemas professed to wanting to increase both attendance and spend per visitor, relying on Virgin's strong brand image and its marketing and retail base.[26] Robert Devereux, chairman of Virgin Entertainment Group, argued that the deal was not about market share but being instrumental in boosting overall admissions from 150 to 200 million per year, helped in part by Virgin's wildly optimistic plans to build 50 new multiplexes in Britain.[27] Virgin's interest did not extend to MGM's non-multiplex sites and in May 1996 they sold 90 of their 119 sites to former Cannon and MGM managing director Barry Jenkins for £68 million, which was raised by venture capital company Cinven. The new circuit was rebranded as ABC in 1996. Virgin retained 18 multiplexes and a range of prime non-multiplex sites, mainly in London.

From the start, it was clear that Virgin saw its multiplex foyers as spaces that were underused and might accommodate a range of ancillary merchandising, including more varied food concessions and CD and video/DVD sales, utilising its Virgin Megastore branding. The aim, according to Devereux, was to "create more traffic through the foyer" and bring customers into the cinema for longer than the running time of the average film and to spend more.[28] Moreover, Virgin was keen to sell and promote its own Virgin Cola soft drink, which had been developed with Canadian company Cott and launched in 1994. A significant innovation was the plan to introduce a monthly pass—the Megapass—designed largely to entice irregular customers back to the cinema, and instal self-service drinks counters and ticket collection machines, all part of a process that was dubbed "Virginalisation" by the company.[29]

Even with a substantially smaller number of sites, Virgin's multiplexes meant that they were the third biggest exhibitor in terms of number of screens and their aggressive expansion, following through in part on MGM's building programme and imminent openings, meant that in the four years after the sale of MGM's non-multiplex sites, Virgin opened 11 multiplexes with 136 screens and in keeping with the wide geographical spread of the MGM circuit, these included sites in Dublin, Rochester, Aberdeen, Birmingham, and Sheffield.[30] Despite Virgin's ambitious and aggressive development programme, based as it was on a desire to increase attendance, in October 1999, less than five years after acquiring MGM, Virgin sold out to French exhibitor UGC (Union Générale Cinématographique) for £215 million. The deal included Virgin's 33 British and two Irish multiplexes totalling 300 screens, along with a further nine sites under planning contract, though it did not include the use of the Virgin brand name.[31] Like so many cases before, the rationale for the sale was in part to do with financial pressures in other parts of the company, with the *Guardian* reporting a leaked memo from Richard Branson concerning a "sudden drop in profits" in its airline operation, along with a forecast that the cinema company would require at least £300 million in capital spending in the following few years.[32] Of course, in many ways Virgin was subject to the demands of its co-investors, who were less confident that attendances, which stood at 139.5 million at the time of the sale, would grow sufficiently to justify the outlay, especially when screen numbers were rising nationally.[33] The takeover of Virgin came just five months before Cinven's acquisition of Odeon, and signalled the beginning of a period of consolidation and rationalisation of the British exhibition market.

UGC had been formed in 1971 as an amalgamation of several circuits and was one of the first major multiplex builders in France, whilst also expanding into neighbouring Belgium (see Chap. 8). In an echo of Branson's "Virginalisation" programme, UGC adopted what it called its "Philosophia" in which the focus was on customer satisfaction and preferences, innovation, and quality.[34] Branson talked up the deal with UGC in terms of his confidence that the new owners would uphold the Virgin ethos. For their part, UGC made the same bold claims that all their predecessors had made upon their entry into the market, when Commercial Director Mike Thomson told *Screen International* that the acquisition of Virgin positioned UGC as "one of Europe's leading cinema circuits—a position from which we aim to build."[35]

One innovation of Virgin's that was retained by UGC was the monthly pass, rebranded as the Unlimited Card, which offered subscribers free entry to any UGC cinema outside central London for a monthly fee of £9.99 (£19.99 for London sites as well as the rest of the country). The Virgin Megapass had cost approximately £15 for four weeks and £25 for eight weeks to all cinemas except Virgin's central London sites; whilst the UGC Unlimited Card was cheaper but required users to sign up for a year. Despite fears that the dominant cinemagoing group—18- to 24-year-olds—would be put off by the credit checks and fact that few had bank accounts from which to deduct the monthly fee, UGC claimed that the Megapass and its replacement were popular—despite the cost of the London pass being reduced to £13.99 in 2002. It is difficult to get a clear sense of the impact of the scheme on admissions; however, Dodona estimated that in terms of income, UGC was receiving approximately £3 per admission; much less than the normal price of a ticket for adults.[36] In any event, as with UGC's rollout of a similar pass in their French multiplexes (see Chap. 9), the introduction of the pass provoked a major row with rival exhibitors and distributors, many of which were smarting from the impact on film rentals of the rapid increase in the number of multiplex screens. Moreover, UGC had been unable to negotiate a drop in film rental prices proportionate to that of the reduced income from admission using the Unlimited Pass. Rival exhibitors felt that distributors were not taking sufficient action against UGC, whilst distributors were of the view that UGC was "cheating them out of the true box office value of their movies, simply to boost its own market share and sell more popcorn."[37] Essentially, the issue was that distributors were unhappy with the cut returned to them by UGC for each Unlimited Card customer. The dispute came to a head when United International Pictures (UIP), which distributed Universal, Paramount, and a range of independent films, levied a top-up of $1.85 on its usual rate of $7.45 on its film *Mission: Impossible II*. UGC claimed that the price increase would compromise its loyalty scheme and despite running trailers for the previous 12 weeks, it threatened to pull the release from its cinemas, before backing down.[38] UGC did issue a writ for compensation claiming breach of contract, which provoked the dispute even more so that in July and August 2000 UIP pulled the releases of both *The Flintstones* and *The Road to Eldorado* from UGC cinemas on the basis, according to UIP, of UGC's "unwillingness to pay a fair rate for the film."[39]

UGC's expansion in Britain was rapid, if short-lived, with five multiplexes opened in 2000 and a further and final four in 2001, including a 4232-seat, 18-screen megaplex in Glasgow's Renfrew Street, which at 62 metres was claimed to be the tallest cinema in the world.[40] In a press release, UGC trumpeted the involvement of Glasgow City Council and the Royal Fine Art Commission in the design of what would be "an attractive landmark,"[41] though even before the building had been completed it was voted "Carbuncle of the Year" in an Internet poll organised by the architectural magazine *Prospect*.[42] UGC then consolidated its presence in Britain and along with its sites in France, Belgium, and other countries in Europe established itself as the second largest exhibitor in Europe. UGC's presence in the British market came to an end in December 2004 when its 42 sites and 408 screens was purchased by the US private equity group Blackstone, owner of Cine-UK, for some £200 million (see below).

## The Multiplex Market Begins to Consolidate

Cineworld was the brand name of Cine-UK, a start-up company formed in 1995 by Steve Wiener, the ex-managing director of Warner Bros Theatres (UK) and claimed to be Britain's first new exhibitor in 40 years.[43] Cine-UK was funded by a group of backers including J.P. Morgan and Rothschild Investment Trust, and the venture capital company Botts & Co which put up the £40 million to start the company, and its initial 14 sites. Wiener and Cine-UK felt that the opportunities for expansion in the British multiplex market were in what they identified as under-screened city centres and out-of-centre developments and smaller catchment areas, which had been hitherto ignored by the larger circuits.[44] Their first site was a 12-screen multiplex on a large leisure park in the new town of Stevenage, opened in July 1996. Its location amidst other leisure attractions including bingo, health club, night club, ten-pin bowling, and restaurants was deliberate, since Cine-UK were keen to capitalise on greater traffic flows rather than risk a stand-alone site. For architectural critic Jonathan Glancey, Stevenage's new cinema was:

> a paradigm of contemporary youth culture, a cinema fuelled on up-to-the-nanosecond screen technology and sited in a sea of edge-of-town burger joints and mall-ish postmodern building, a perfect suburban frame for summertime gum-chewing, sulking, hair-flicking, mountain-bike wheelies and the wearing of baseball caps with the peaks at the back.[45]

The most significant development, which would come to symbolise many of Cine-UK's subsequent builds, was its compact design and footprint. Stevenage's 12 screens occupied 36,500 square feet at a time when a ten-screen complex was routinely occupying some 40,000 square feet, which it managed by scaling down nine of the auditoria whilst claiming to maintain the ratio of screen size to seating space.[46] Stevenage heralded a rapid expansion of Cineworld sites, which by the end of 2000 numbered 19 with 210 screens. Many of these were in town centre locations, such as Wakefield, Chesterfield, Shrewsbury, Luton, Weymouth, and Burton-on-Trent and/or in edge-of-centre leisure parks near towns such as Wolverhampton, Swindon, Ashford, and Runcorn. One of the innovative aspects of Cine-UK's approach, in part a result of some of its locations, was its programming of Hindi-language or "Bollywood" films under standard profit-sharing distribution terms. The company had piloted the screenings at its four-screen Feltham site in London before rolling them out to venues like Ilford, Luton, Bradford, and Wolverhampton. By 2007, Cine-UK had 55 per cent of British market for Hindi-language films.[47] At their giant StarCity site (see below), Warner Village dedicated six screens to Hindi-language films, arguing that in part it was commercially viable because of a clampdown on pirated videos.[48]

Cine-UK also set in train an ambitious expansion, with 16 multiplexes proposed for 2001–02 and like their first tranche these were in what many analysts called "secondary markets"—smaller towns and regional centres. In reality, Cine-UK had opened ten Cineworld multiplexes by the end of 2002 and a further three by the end of 2003, whereupon Cine-UK's 32 multiplexes (357 screens) had made it the fourth largest cinema circuit in Britain.[49] Amidst much speculation about the future of the company as a target for consolidation, it was acquired by US private equity group Blackstone in September 2004, for approximately £120 million. The sale of Cine-UK can be put in context by the circuit's head Steve Wiener, who said that "the original business plan had been to open eight cinemas and then sell, but by the time we had six operational it was clear that we were on to something."[50]

Cine-UK's new owners, Blackstone, then purchased UGC in December 2004 and merged the two circuits, rebranding all of UGC's sites as Cineworld. This made the circuit the second largest in Britain with 20–25 per cent of the market by numbers of screens, seats, and box-office and, as in the purchase of Odeon/UCI by Terra Firma, it attracted the attention of the Office of Fair Trading (OFT).[51] Similarly, the OFT judged that the merger of Cine-UK and UGC should be referred to the Competition Commission unless the new company addressed the competition concerns. In order to stave off the Competition Commission the merged company agreed to divest itself of cinemas in local areas where prior competition between Cine-UK and UGC existed, including: Swindon, Wigan, and Birmingham Great Park. These cinemas, along with four others, were acquired by a new exhibitor called Empire, which was owned by the Irish exhibitor Ward Anderson. Empire had also purchased 11 sites from Odeon/UCI, which had also been forced to sell sites by the OFT as a result of that merger.

Having been floated on the stock exchange and renamed Cineworld plc., the company would attract the attention of the OFT again in 2013 having acquired City Screen Limited for £47.3 million, which operated 21 cinemas under its Picturehouse brand, in December 2012. City Screen had been formed in 1989 and sought to offer its audiences a broader mix of films, including so-called arthouse, in addition to operating in city and town centre locations. The OFT concluded, however, that since the cinema market was "generally acknowledged in the industry as being of three main types—multiplex, art-house and luxury" then as a multiplex operator, Cineworld did not compete with Picturehouse, as an arthouse operator.[52] Indeed, Cineworld continued to operate Picturehouse as a separate entity, maintaining its identity as an exhibitor whose audience was different from that of the parent company. Cineworld's Philip Bowcock promised that they would not come in "as a big corporate giant saying 'you're going to do xyz'. We're going to keep the quirkiness. It's a very different customer set—a little bit older, more discerning, more experienced. To lose that would be to lose the raison d'être of Picturehouse."[53]

In Britain, Warner Bros. had opened their only cinema—the Warner West End—in 1938, but in 1988 Warner Bros. set up a new division—

Warner Bros. International Theaters (WBIT)—to manage its expansion into international exhibition, for which Britain, with its rapidly expanding multiplex sector, was of major significance. Shortly after the formation of their new international exhibition division, the company announced an initial slate of eight multiplexes in Britain, including sites in Newcastle, York, Basingstoke, and Preston. The first to be opened though was in Bury near Manchester in June 1989, which had 12 screens and seated over 3900 people. The site chosen was part of a leisure park adjoining a large supermarket, a ten-pin bowling alley, and a range of restaurants including McDonald's. In contrast to Cineplex-Odeon's focus on the area around London, WBIT opted for locations like Bury because they were cheaper and more accessible. Interviewed in *BoxOffice*, WBIT president Millard Ochs talked of the move into the British market as "a mission" based on the poor exhibition infrastructure but a sense that people would return to the cinema if offered "a whole new moviegoing experience."[54] As a vertically integrated company Warner Bros.' film distribution arm had the third largest share of the British market in 1988 with 20 per cent,[55] whilst its video distribution arm had the largest share of rental transactions—16 per cent in 1988.[56] Many in the industry were of the view that WBIT's strategy for its cinemas was historically to sell them some time after opening them, since they saw their exhibition interests acting as a "catalyst to expand exhibition markets to the benefit of [their] film production and distribution interests."[57]

Unlike National Amusement's Showcase circuit, WBIT initially lacked what MEDIA Salles called a "clear concept for their sites" since the number of screens across their first eight multiplexes varied widely from complex to complex.[58] Bury and York had twelve screens, Newcastle had nine screens, and both Thurrock and Preston had seven screens. Nevertheless, though a late entrant to the British exhibition market, WBIT began an accelerated opening policy so that by 1996, five years after their first opening, they had 19 sites with 116 screens.[59] In October of that year, WBIT formed a joint venture with Australian exhibitor Village Roadshow International called Warner Village Cinemas Ltd, to own and manage its cinema operations in both Britain and Germany. It was an extension effectively of a pre-existing venture in Australia that went back to the early 1970s (see Chap. 10) and meant that WBIT's sites were rebranded as Warner Village Cinemas. The merger of the more conservative WBIT and more aggressive Village Roadshow meant that the company expanded a number of sites and began building larger multiplexes, including a

15-screen complex in Plymouth. By December 1999, the company had 27 sites with 261 screens.

## HERE COMES THE MEGAPLEX—"THE MOVIE EXPERIENCE FOR THE MILLENNIUM"

Having built one large multiplex, Warner Village opened a gigantic 30-screen cinema in Birmingham's StarCity in July 2000, which seated 6200 people and was then Europe's largest cinema.[60] The development at StarCity was significant because it signalled the new generation of mega-plexes—which the Union Internationales des Cinémas (a trade body representing exhibitors across Europe) defined as a cinema with at least 16 screens. Warner Village made great play of their intentions to open several across Britain, including the much vaunted but never realised "Battersea Powerplex," which would have been a 24-hour, 32-screen in the derelict power station on the bank of the River Thames in London.

Virgin had registered the term "Megaplex" with the UK *Trade Marks Journal* in July 1997 and insisted that it had sole rights to use the term; which it applied to its 20-screen, 4821-seat complex in Sheffield, opened in November 1998, and which was briefly the largest cinema in Britain. Copyright notwithstanding, this had not prevented UCI from claiming that its 20-screen site in Manchester's Trafford Park was Britain's first megaplex.[61] Moreover, Warner Village were also using it to describe StarCity. There ensued a legal war of words in which Warner Village claimed that "megaplex" was a "descriptive word that cannot be enforced. Virgin didn't coin it as it was already in the public domain" whilst UCI said that "Virgin's registration is a sham."[62] For Virgin, the prefix "mega" was as much part of their branding (e.g. "Megastore") and in the new 13-screen site in Great Park in Birmingham, though small by US and European megaplex standards, signalled a new trend in larger multiplexes. The site was also the first in which all of the screens were installed with THX certificated sound, which formed the centrepiece of Virgin's promotional material upon opening.[63] In addition, Great Park was one of the first sites with what was called "Premier Screen"—a premium-priced smaller auditorium—described as the "the cinema equivalent of flying 'upper class' with its own private bar and waitress service."[64] Given that Sheffield already had three multiplex cinemas, Dodona was of the view that the new 20-screen Virgin megaplex was intended as what it called a "category killer cinema" in that it contained two Premier Screens, a restaurant and café, a party room, and a CD and memorabilia store.[65] This move to a differentiated screening and pricing policy was imitated by other exhibitors which were keen to stress the importance of 'customer care,' 'value for money,' and quality assurances with regard to facilities. The Warner Village megaplex in StarCity gave over three screens to what it called Gold Class, in which cinemagoers sat in reclining leather armchairs, consumed complimentary popcorn, and were able to purchase alcohol.

Warner Village undoubtedly gambled on spending some £15 million on the 135,000-square-foot StarCity complex, which they claimed represented nothing less than the "the re-invention of the cinema-going experience in the UK."[66] The cinema was the anchor tenant in a 50,000-square-foot shopping, leisure, and restaurant complex—characterised by Wilkinson as "an outpost of America just off Spaghetti Junction"—on the site of a former gas works in the Heartlands area of the city.[67] StarCity marked another development in cinema design copied from the USA, with the complex's US architect Jon Jerde noted for his flamboyant shopping malls. In StarCity, the building's aesthetic was described by the architect as a "new generation of urban entertainment" drawing its inspiration from "Birmingham's rich tradition as the birthplace of the industrial revolution."[68] In a promotional feature in 1998, Warner Village set out its strategy for the future, with its multiplexes augmented by what it called "the giant multiplex" being built at StarCity: the "movie experience for the millennium."[69] "Operations-wise," Warner Village stated, "the Birmingham 'plex is really just an oversize multiplex yet it is simply not enough to offer layer upon layer of screens as they were nothing more than auditorium wallpaper ... [the] objective is to customise the screens and create a variety of cinematic destinations under one roof."[70]

The financial projections for the cinema were indeed optimistic with 1.3 million visits required just to break even, prompting Warner Village's managing director to admit to the *Financial Times* that "from one point of view, it's a crazy idea."[71] Two weeks before StarCity began operation, UGC opened its flagship 12-screen, 2952-seat multiplex in Broad Street in Birmingham's city centre, meaning that overnight the number of screens in the city doubled. Despite a very impressive opening week when it took £150,000, Warner Village was sanguine about the reality that making back its huge building costs would take many years, even with predictions of 30–40 per cent box-office growth in the following year.[72] The company's anxieties were demonstrated when in 2002, with competition from a growing number of multiplexes opening in Birmingham, including AMC's return to the British market with a 12-screen multiplex near UGC's Broad Street site, Warner Village undertook a rationalisation of the StarCity site and closed nine of the 30 screens, removing all fixtures and equipment.

## The Emergence of a British 'Big Three'

The following year, in May 2003, Warner Village Cinemas Ltd sold its 36 multiplexes to SBC International Cinemas for £250 million. SBC (Spean Bridge Cinemas), which had been formed in 1998 by Tim Richards, a former employee of WBIT and Stewart Blair, operated three multiplexes in Scotland (Livingstone, Hamilton, and Aberdeen) and one in England (Southport). Its acquisition of Warner Village was backed by a number of private equity groups, including Boston Ventures from the USA and Legal and General from Britain, and also saw the company's name change to Vue Entertainment Holdings (UK) Ltd. Vue set about rebranding the cinemas the following year.

Vue's head Tim Richards inherited a largely modern circuit of 45 multiplexes, 38 of which were built since 1995 and had modern stadium seating, though he professed to wanting Vue to appeal to an audience beyond simply that of 15- to 24-year-olds. To this end, Vue began revamping the former Warner Village sites, removing the cartoon character iconography and "discotheque aesthetics" and redesigning the interiors with more muted colours and including coffee bars, whilst the film programming was diversified to include films more attractive to an older audience.[73] In order to reach potential older audiences Vue utilised a profiling system called MOSAIC in order to analyse consumer trends. The acquisition of Warner Village meant that Vue was a major player from the start and adopted a twin policy of growing through acquisition and via an ambitious building programme.

In terms of acquisition, Vue bought US company Cinemark's two sites in Northampton and Scunthorpe in 2004, followed in 2005 by the purchase of Ster Century—the brand name for the international operations of South African company Ster-Kinekor—for £50 million and which operated six multiplexes in Britain including sites in Dublin, Norwich, Leeds, and Cardiff.[74] In 2012, Vue acquired Apollo Cinemas' 14 small multiplexes, most of which were in smaller towns across Britain, though again, competition issues mean that they were forced to divest themselves of four of these to small rival Reel Cinemas. Vue's expansion via construction saw it increase its screen count year on year so that by 2010, when Vue was sold to private equity firm Doughty Hanson & Co. for $730 million, the company had doubled its screen count from 385 to 655, with a market share of 22.7 per cent of Britain box-office.[75] In 2012, Vue began to look towards mainland Europe for expansion opportunities and purchased a majority stake in CinemaxX, the second largest circuit in Germany (see Chap. 9), and Multikino, the second largest circuit in Poland.

The takeovers and mergers of Odeon/UCI, Cine-UK/UGC, and Vue/Warner Village meant that by the end of 2005, the industry had entered a period of consolidation as companies began to increase their circuits via a process of acquisition. One can get some sense of this by considering that, in the 15 years of its operation, Britain's second multiplex, opened in Salford Quays by Cannon Cinemas (though planned by Thorn-EMI) in December 1986, had five different exhibitors' names above its doors: in 1990 Cannon was taken over by MGM; in 1995 it was acquired by Virgin; it was taken over by UGC in 1999 and finally became a Cineworld, which closed it in 2001.[76] Many of these major takeovers demonstrated the renewed interest in the cinema sector by private equity companies, drawn perhaps by the possible real estate values of many cinemas, particularly those located in city centres. According to Dodona Research, in the two years prior to March 2005 some two thirds of cinema screens in Britain had been traded, meaning that the multiplex industry was dominated by three companies—Terra Firma (UCI and Odeon), Cine-UK (Cineworld), and Vue—which controlled 61.1 per cent of screens in Britain.[77]

Having diversified considerably after the duopoly of the post-war period, this new spate of mergers and takeovers saw the exhibition sector once again consolidated to a significant degree. More significantly, AMC, which left the British market in the late 1980s, re-entered with two multiplexes in Manchester and Birmingham in 2002, before purchasing Odeon/UCI and is the last remaining representative of the several foreign companies which entered the market in the late 1990s and early 2000s. All the rest, Village Roadshow, CIC, Hoyts, Cinemark, Ster-Kinekor, and Cineplex Odeon were acquired at various points by the three leading companies.

## NOTES

1. See Stuart Hanson, "The Rise of the Multiplex," in *The Routledge Companion to British Cinema History*, eds. I.Q. Hunter, Laraine Porter, and Justin Smith (London: Routledge, 2017).
2. "United Kingdom Cinema Market," *Screen Digest* (July, 1991), 158.
3. Dodona, *Cinemagoing 9* (Leicester: Dodona Research, 2001), 47.
4. Dodona, *Cinemagoing 15* (Leicester: Dodona Research, 2006), 23.
5. "Rank Aims for 200 Odeon Screens in 75 Locations with Expansion Plans," *Screen International* 601 (23 May, 1987), 22.
6. Dodona, *Cinemagoing 4* (Leicester: Dodona Research, 1995), 29.
7. Ibid., 29.
8. In 1994, Paramount Communications was purchased by Viacom International, which was owned by Sumner Redstone's National Amusements. Universal was purchased by the French company Vivendi in 2000 to create Vivendi Universal.
9. It also managed two central London sites, the Empire and The Plaza, for CIC.
10. Dodona, *Cinemagoing 9*, 64.
11. "UCI Opening 8-plex, London's First Multi," *Variety* 337, no. 7 (22 November, 1989), 17.
12. "Cannon to Close Seven City Screens," **Screen International** 661 (16 July, 1988), 2.
13. Celia Duncan, "10 Years of UCI: If You Build It, They Will Come," *Screen International* 1026 (22 September, 1995), 28.
14. Office of Fair Trading, *Acquisition by Terra Firma Investments GP. 2 Ltd of United Cinemas International UK. Limited and Cinema International Corporation UK. Limited*, 7 January 2005. https://assets.publishing.service.gov.uk/media/555de40540f0b669c40000fb/terra.pdf (accessed 8 August, 2016).
15. Ibid., para. 39.
16. Dodona, *Cinemagoing 12* (Leicester: Dodona Research, 2004).
17. Joseph Cotterill and Murad Ahmed, "AMC to buy Odeon from Terra Firma for £921m," *Financial Times*, 12 July, 2016.
18. See Robert Marich, "Cannon Special: Focus Cannon Group," *Variety* 332, no. 11 (5 October, 1998): 49–51.
19. Philip Turner, *Cannon Cinemas: An Outline History* (St. Paul's Cray: Brantwood Books, 1997).
20. See Turner, *MGM Cinemas*.
21. "Multiplex Cinemas: Growth is Far from Over," *Screen Digest* (February, 1995), 36–7.

22. Ralf Ludemann, "Exhibition," *Screen International* 891 (22 January, 1993), 13.
23. Monopolies and Mergers Commission, *Films: A Report on the Supply of Films for Exhibition in Cinemas in the UK*, Cm 2673 (1994).
24. Ibid., para. 1.17, 5.
25. For the most detailed history of Virgin's multiplex circuit, see Allen Eyles, "The Virgin Intervention," *Picture House: Magazine of the Cinema Theatre Association*, no. 43 (2018): 8–38.
26. Jonathan Davis and Neil Reeder, "Chain Reaction," *Screen International* 1019 (4 August, 1995), 12–13.
27. Beth Porter, "Virgin Exhibitors," *The Film Journal* 98, no. 9 (1 December, 1995), 16.
28. Ralf Ludemann, "Interview Robert Devereux: Empire Builder," *Screen International* 1019 (3 August, 1995), 14.
29. Ralf Ludemann, "Virgin Territory," *Screen International* 1015 (7 July, 1995), 10.
30. Dodona, *Cinemagoing 9*.
31. Mary Scott, "Virgin Pulls UK-Irish Exhib Plug," *Screen International* 1231 (22 October, 1999), 1.
32. Lisa Buckingham, "Virgin Sells 'Fantastic' Cinemas for £215m," *The Guardian*, 19 October, 1999.
33. Ibid.
34. Fabrizio Montanari, *When the Audience Comes First: The Case of UGC Cinemas Institute of Organization and Information Systems* (Milan: Bocconi University, 2003). http://ernest.hec.ca/video/pedagogie/gestion_des_arts/AIMAC/2003/resources/marketing.htm (accessed 11 August, 2016).
35. Scott, "Size Matters in UK Exhib'n Race," 8.
36. Dodona, *Cinemagoing 11* (Leicester: Dodona Research, 2003), 47.
37. Adam Dawtrey, "UGC Pass Piques Plex Players," *Variety* 379, no. 8 (17 July, 2000), 66.
38. Stephen W. Gilbert, "Foreign Correspondence: The British Picture," *Film Journal International* 103, no. 10 (1 October, 2000), 84–5.
39. Mary Scott, "UGC Card Provokes UK Distribution Fury," *Screen International* 1270 (4 August, 2000), 4.
40. Denise Roland, "Cineworld and Its Founder: Nine Things You Did Not Know," *The Daily Telegraph*, 20 November, 2013. https://www.telegraph.co.uk/finance/newsbysector/retailandconsumer/leisure/10462630/Cineworld-and-its-founder-nine-things-you-did-not-know.html (accessed 13 November, 2017).
41. UGC, "UGC to Launch Three New Cinemas in Three Months" (UGC Cinemas: London, n.d.).

42. *BBC News*, "Cinema 'Sight for Sore Eyes'," http://news.bbc.co.uk/1/hi/scotland/1028193.stm (accessed 24 November, 2018).
43. See Ralf Ludemann, "Birth of a Notion," *Screen International* 1065 (5 July, 1996): 15–16, 19–21.
44. See Ralf Ludemann, "Cine UK Unveils Exhibition Programme," *Screen International* 1011 (9 June, 1995), 1 and Beth Porter, "Cine-UK's Wiener Challenges Big Exhibition Guns," *The Film Journal* 99, no. 1 (1 January, 1996), 10, 110.
45. Jonathan Glancey, "Slouching towards Stevenage," *The Independent Magazine*, 7 July, 1996, 20.
46. Ludemann, "Birth of a Notion," 18.
47. Charles Gant, "Passage from India," *Sight and Sound* 17, no. 10 (October, 2007), 9.
48. Angelique Chrisafis, "Hoorah for Bollywood as Birmingham Gets Multiplex Screens Dedicated to Indian films," *The Guardian*, 17 May, 2000.
49. Dodona, *Cinemagoing 14* (Leicester: Dodona Research, 2005), 37–8.
50. Quoted in Mark Moran, "Cinema showman," *Cinemabusiness* 28 (September, 2006), 18.
51. Office of Fair Trading, *Completed Acquisition by the Blackstone Group of UGC Cinemas Holdings Limited*, 5 May, 2005. https://assets.digital.cabinet-office.gov.uk/media/555de43240f0b666a2000100/ugc.pdf (accessed 14 January, 2017).
52. Office of Fair Trading, *Completed Acquisition by Cineworld plc of City Screen Limited*, ME/5877/12. 5 June 2013, Para. 3: 13.
53. Quoted in Julia Kollewe, "Cineworld Buys Picturehouse," *The Guardian*, 6 December, 2012.
54. Bridget Byrne, "Exhibition Profile: Making WBIT Tracks: Warner Bros. International Theatres Celebrates Its First Decade," *BoxOffice* 134, no. 6 (1 June, 1998), 40.
55. "British Cinema and Film Statistics," *Screen Digest* (October, 1990), 14.
56. Jeremy Coopman, "HV Loses No Luster in U.K.; Yanks Clean Up on Rental," *Variety* 335, no. 3 (10 May, 1989), 187.
57. Dodona, *Cinemagoing 11* (Leicester: Dodona Research, 2003), 51.
58. MEDIA Salles. *White Book of the European Exhibition Industry: A Report by London Economics and BIPE Conseil: Vol 2*, (1994) http://www.mediasalles.it/whiteboo/wb2_1_3.htm (accessed 4 January, 2019).
59. "Multiplex Cinemas: Growth Is Far from Over," 37.
60. See Turner. *Warner Cinemas.*
61. Scott, "Size Matters in UK Exhib'n Race," 8.
62. Ibid., 8.

63. "Virgin Territory," *Film Journal International* 101, no. 7 (1 July, 1998), 26, 28.
64. Ralf Ludemann, "UK Preview 1997: Battle of the Plexes," *Screen International* 1092 (24 January, 1997), 33.
65. Dodona, *Cinemagoing Europe 1999* (Leicester: Dodona Research, 1999), 45–6.
66. "Warner Village's Rising Star," *Screen International* 1269 (28 July, 2000), 14.
67. Carl Wilkinson, "Welcome to Cinema Paradiso," *The Observer*, 17 August, 2003.
68. Jon Verde quoted in Gary Younge, "The Big Picture," *The Guardian*, 26 July, 2000.
69. "10 years of Warner Bros. International Theatres: Promotional Feature," *Screen International* 1150 (20 March, 1998), 28.
70. Ibid., 28.
71. Jonathon Guthrie, "Stars bring Beverly Hills to Birmingham," *Financial Times*, 21 July, 2000.
72. Mary Scott, "StarCity Highlights Megaplex Concerns," *Screen International* 1269 (28 July, 2000), 2.
73. "The Operators," *Cinemabusiness*, December, 2005, 18–20.
74. The acquisition of Ster Century was referred to the OFT, who ruled that as Vue had a near-monopoly situation in Basingstoke the company had to sell its Basingstoke Leisure Park site to rival Odeon. See *Screen Digest*, "Movie Markets in the UK A report commissioned by OFCOM," December, 2007. https://www.ofcom.org.uk/__data/assets/pdf_file/0020/54605/annex11.pdf (accessed 24 July, 2017).
75. Dodona. *Cinemagoing 20* (Leicester: Dodona Research, 2011), 19.
76. Hanson, "The rise of the multiplex."
77. UK Film Council, *Statistical Yearbook 2005/06* (London: UK Film Council, 2006), 51.

# CHAPTER 7

# Multiplexes Move Out of Town and Back Again

The diffusion of the multiplex in Britain echoed that of the USA, where the development of the multiplex from the 1960s was inextricably linked to the importance of the suburban, out-of-town shopping centre (or mall). In Britain, correspondingly the growth of the multiplex would take place in out-of-town shopping and leisure complexes around many major conurbations, the demand for which was related to the steady increase in car ownership throughout the 1980 and 1990s and the move by many people to housing developments on the urban fringe. The emergence of a new kind of out-of-town leisure and shopping culture in the 1980s and 1990s was both a result of economic and political changes, and indicative of a new set of aesthetics around shopping and consumerism. Of crucial importance was the role of urban planning and in particular the easing of planning restrictions in the 1980s, with the concomitant development of large out-of-town shopping centres.[1] In the 1980s, a series of regeneration policies by central government sought to speed up the planning process and create a relaxed regulatory environment in order to stimulate commercial and retail development. As Britain's traditional manufacturing industries declined, the emphasis was increasingly placed on the rapid development of out-of-town shopping and leisure complexes across Britain, often on old industrial sites. In line with domestic developments, US multiplex owners continued to look to Britain for new sites for their new cinemas.

© The Author(s) 2019
S. Hanson, *Screening the World*,
https://doi.org/10.1007/978-3-030-18995-2_7

## THE MULTIPLEX CINEMA OUT OF TOWN—"OKAY FOR VOLVO OWNERS"

By the early 1990s, it is no exaggeration to suggest that Britain had become the most developed multiplex market in Europe. Within five years of the opening of The Point in 1985, 42 multiplexes were built in what can only be described as a building boom. This was surpassed in a second boom from 1990 to 1995 when a further 40 multiplexes were constructed, so that in the first ten years 82 multiplexes were built in areas carefully chosen on a range of criteria, including the multiplex's accessibility to the surrounding population. The imperative on the part of the multiplex operators was to make it convenient for the surrounding population to drive to the cinema, park safely and for free. The sites chosen were overwhelmingly, though not exclusively, in out-of-town locations around major cities and conurbations.

In 2003, *Screen Digest* reported that the proportion of cinema screens in Britain's multiplexes was the highest in Europe at 67 per cent; a position Britain had maintained since 1998, when it was the only country in Europe with more than half of its screens in multiplexes.[2] Moreover, in 2002, the frequency of cinemagoing in Britain reached 2.98 times per capita: the highest since 1971 as 176 million visits were made to the cinema.[3] By 2017, 78.2 per cent of screens were in multiplexes, though the frequency of visits was down to 2.6 times per capita.[4]

In the USA, as we have seen, the period after the Second World War witnessed a large-scale movement of people from out the city into the burgeoning suburbs, and, whilst incomparable in terms of the numbers, Britain's large cities also saw significant movements. Post-war planning policy in Britain was largely concerned with what Wannop and Cherry identified as "a basic strategy for population decentralization with a following movement of industry, backed by measures to more sharply differentiate town from country."[5] This entailed the movement of populations in Britain's conurbations to new communities, notably in new towns and the developing suburbs. In part, this was determined by the consequences of German bombing, which damaged or destroyed some 3.75 million houses in Britain.[6] In addition, there existed in many towns and cities a large stock of substandard housing, many of which were defined as slums. In London alone in 1949, 20 per cent of homes were officially classed as slums, whilst by 1951, eight million homes in Britain had been declared unfit for habitation.[7] In England and Wales between 1955 and 1985, 1.48 million houses were demolished and 3.66 million people displaced from city centres.[8] Thus, it was the movement of largely working-class communities out of the city, usually as part of slum clearance programmes, to new suburbs and towns on the periphery of the conurbation which played a significant part in the decline of cinemagoing.[9]

The focus of developments in the new re-built and re-shaped urban centres themselves was concerned with the commercial centre, which witnessed widespread re-building of shopping and business facilities in the 1950s and early 1960s. However, this was rarely accompanied by cinema development, especially outside London, and many cinemas destroyed by wartime bombing had not been replaced. For some ten years after the end of the Second World War, there were restrictions on luxury building, with housing the priority for planners. This meant that, notwithstanding some new cinemas built in the 1960s to replace some of those destroyed by bombing, there was no significant cinema building in Britain until the multiplex in 1985.[10] Nevertheless, the city and town centres still contained many of the largest and most prestigious cinemas; therefore, the major exhibitors concentrated on the modernisation of these sites, including the dividing up of cinemas into multiscreen venues. This meant that the exhibitors' economic futures were tied to a large extent to those of the city, though in the context of a steadily declining population in the city centre from the 1950s until the 1990s.

The fortunes of the city and town centres were linked to the issue of whether retail development should be allowed 'out of town' as distinct from 'in town' and this was the key issue that defined planning policy in Britain.[11] This had resulted in a series of tensions amongst urban planners and developers since the 1970s as Britain's provincial inner cities had declined in relative importance as commercial and retail centres. The turning point was in part, according to Schiller, the development of large out-of-town shopping centres, which he characterised as the "third wave" of retail decentralisation in Britain.[12] The 'first wave,' in the 1970s, had seen the development of large supermarkets outside of town centres, followed by the 'second wave' of what were called retail parks primarily involving DIY centres, electrical retailers, and furniture stores (i.e. IKEA) from the mid-1980s onwards. The growth of these retail parks was, according to Schiller, "a phenomenon with no direct parallel in other countries."[13]

This trend towards out-of-town developments in Britain was part of a general shift in favour of shopping and leisure in what was rapidly becoming a post-industrial economy, allied to the on-going growth in car ownership. There was a steady increase in car ownership throughout the 1980s, which has continued through to 2017. In 1985–6 the proportion of households with access to a car was 62 per cent, rising to 70 per cent in 1995–7 and then to 76 per cent in 2017.[14] In 2000, the number of households with two or more cars exceeded the number of households with no car.[15] Throughout the post-war period and especially into the 1980s, many people had moved to housing developments on the urban fringe, which led to greater demands for out-of-town shopping facilities. Moreover, many retailers and particularly the large circuits sought to reduce rents and combat the threat to business from city centre congestion by moving outwards.

During the 1980s, the Conservative government under Prime Minister Margaret Thatcher had sought to encourage inward investment via a relaxed regulatory culture, in which planning authorities had been instructed to be less rigid when considering applications for planning consent, in line with the government's free enterprise philosophy.[16] This much more *laissez-faire* approach to regional planning meant that there would be a presumption in favour of developments such as shopping centres.[17] The provision of new shopping facilities was a key element in the planned regeneration of regional economies, in particular older industrial areas that had effectively become de-industrialised. The government proposed that the regeneration of these areas would come as a result of providing new jobs and better living conditions. This would not be achieved solely by direct government investment but would require private capital. The imperative was to stimulate the local economy and provide employment.

For many of the developers, this inevitably meant growth of the service sector in the form of retailing and leisure-based industries. Essentially, the Thatcher government's streamlining of planning procedures was a consolidation of the role of planning in supporting the market; indeed, it was nothing less than an effort, Griffiths argued, to "restructure for the market."[18] Some sense of the deeply ideological tenure of the government's proposals can be found in successive legislative changes to planning throughout the 1980s. In 1980, the Local Government, Planning and Land Act introduced Enterprise Zones and Urban Development Corporations. In the Enterprise Zones planning controls were relaxed with certain kinds of development given what Guy argued was "in effect automatic planning permission."[19] Like Urban Development Corporations, which were created in order to regenerate former industrial areas, Enterprise Zones relied on private investment attracted by this relaxed planning regime and also by a series of financial incentives. That *ad hoc* retail and leisure developments would dominate these areas was the outcome of a system in which no reference needed be made to a formal plan.[20] In 1985, the Housing and Town Planning Bill introduced what were called Simplified Planning Zones, in which, as their name suggests, planning was granted in advance for particular type of development. Like previous changes in legislation, the emphasis was on wealth creation and regeneration, with the private sector providing the majority of the capital.

The result of such shifts in planning in the 1980s and early 1990s was that property markets became more speculative and the drive on part of retailers to get more efficient returns on capital meant that most proposals for major out-of-town shopping centres were approved, usually on the outskirts of major conurbations near motorway junctions or major orbital roads. US cinema companies saw the parallel between these developments and those that had been so successful in the USA, especially since many catered for the motorcar. This meant that the development of the multiplex cinema in Britain was intimately bound up with Schiller's "third wave" of retail development and decentralisation.[21] A salient example of this phenomena was Britain's second multiplex cinema opened in Salford Quays in 1986; the former Manchester Docks and designated an Enterprise Zone in 1981. The multiplex had been planned by Thorn-EMI Screen Entertainment (TESE)[22] whose ambitious head, Gary Dartnall, recognised the potential of new-build, multiscreen cinemas. On announcing the new multiplex in Salford Quays, Dartnall described cinema exhibition as:

> a neglected industry which has failed to follow the audience from inner cities out to the suburbs. Modern cinema-goers want somewhere to park, somewhere to eat and pleasant surroundings, which we intend to give them in Salford ...We are doing what the Americans did as long ago as the 1960s— making cinemagoing an event—and look how healthy audiences are over there.[23]

In selecting Salford Quays TESE were able to take advantage of the area's designation as an Enterprise Zone, which was exempt from Development Land Tax and council rates, and subject to simplified planning and controls.

Between 1986 and 1990 four giant out-of-town regional shopping centres were opened—Metro Centre near Gateshead/Newcastle-on-Tyne; Merry Hill on the site of a former steelworks in the industrial West

Midlands near Dudley; Lakeside near Thurrock on the M25 London orbital motorway; and Meadowhall near Sheffield, also on the site of a former steelworks. A regional shopping centre was one that that contained greater than 50,000 square metres (540,000 square feet) of retail and leisure space and away from a town or city centre.[24] These centres were the location for the first round of multiplex developments. On three of these sites, American Multi-Cinema (AMC) and later United Cinemas International (UCI)[25] opened some of the first multiplexes in Britain, part of the aforementioned boom in multiplex building which took place between 1985 and 1990; whilst Warner Bros. opened one of its first multiplexes at Sheffield's Meadowhall in 1993. Edward Soja, in his observations on Los Angeles, described the urban sprawl and absence of a core as an *exopolis*, meaning "the city without."[26] In this conception, the urban sprawl is an aggregation of various centres in which, as Phillips argues, the "concept of metropolitan areas as an urban core with dependent suburbs is outdated."[27] Utilising journalist Joel Garreau's conception of the "Edge City,"[28] Barker referred to a process whereby for the preceding ten years Britain's cities had been "turned inside out."[29] Focussing on the proposed Bluewater shopping centre,[30] which was being constructed on a former quarry, Barker observed:

> If you want a 20/20 vision of the Edge City future, go to Dartford in Kent. This is motorway country—the M25, M2, M20. It is also quarry country: a chalky, lunar landscape of abandoned cement workings … It is notable how many of the Conservatives' regeneration schemes end up pushing the Edge City phenomenon forward.[31]

For the foreseeable future, one trade journal argued, the fate of the multiplex would "be bound up with that of the shopping centre."[32]

The multiplex construction that was undertaken in the initial building phase from 1985 to 1990 took place in areas carefully chosen using a range of criteria, including the multiplex's accessibility to the surrounding population. In Britain, this figure was minimally 200–300,000 people within a 20-minute drive,[33] up to 500–600,000 people within a 45-minute drive.[34] The imperative on the part of the multiplex operators was to make it convenient for the surrounding population to drive to the cinema and park safely, conveniently and for free. The decision to build in any location was determined by a complex set of issues around potential demand as compared to the cost base such as rent, depreciation, maintenance, and utilities.[35] The sites chosen were overwhelmingly, though not exclusively, in out-of-town locations around major cities and conurbations. Very often these were greenfield sites, cited frequently by developers as the only places on which multiplexes could be constructed, since with anything up to 12 screens and auditoria on a single floor the new multiplexes needed large ground areas of 3000–4000 square metres, which along with car parking added up to some 15,000 square metres.[36]

A reliance on private, rather than public, transport was the corollary of these out-of-town developments and this mobile population was one that the multiplex was able to court since they could offer sites that were convenient, near motorways, with free car parks, whilst many city cinemas could not. In their study of the patronage of multiplex and non-multiplex cinemas, Collins et al. argued that those who travelled to out-of-town multiplexes were less sensitive to travel time and distance than those who went to cinemas in city centres.[37] In part, this might reflect a reliance on and necessity for private transport on the part of those who lived outside of city centres. This hypothesis seemed borne out by the Cinema Advertising Association's (CAA) 2002 *Cinema and Video Industry Audience Research (CAVIAR) 20* audience study, which indicated that 74 per cent of people who visited the cinema travelled there by car.[38] Speaking at a seminar organised by the British Film Institute (BFI) in 1985, Pamela Hare-Duke, the Cinema Director of the Cornerhouse in Manchester, criticised the then-current plan "to build multiplexes at the end of motorways," adding that they would be "okay for Volvo owners but not for viewers."[39]

Retail and leisure developments throughout the 1980s and early 1990s, then, were dominated to a large extent by: an increase in car ownership; the long and steady process of urban decentralisation; a lack of suitable sites for development in city centres; cheaper land and property rents in out-of-town sites; and the introduction of development zones with the concomitant relaxation in planning. The accompanying increase in multiplex developments proceeded apace and notwithstanding some slowdowns in openings and the realisation that in some regions competition between multiplexes had resulted in over-screening; the majority of sites opened were outside of the traditional city centres. Dodona was unambiguous in its assessment of cinema closures, stating that they "generally represent a contribution to the process of moving cinemas from where they are not wanted to where they are, most often to somewhere with parking."[40] It was "no coincidence," Wroe observed, that "British leisure is joining British retailing in an exodus from the town centre to its edges."[41]

## THE MULTIPLEX CINEMA—FROM GREENFIELD
## TO BROWNFIELD

As we have seen above, the development of the multiplex in Britain, particularly in the first ten years, was inextricably linked to the development of the large out-of-town shopping centres. By the early 1990s, however, the consequences of out-of-town centres in terms of their impact on city centres seemed clear to many city councils, which, in the words of AMC's Bruno Frydman, feared that the advent of multiplexes would take the cinemagoing public away from town and city centres, with audiences irresistibly drawn by the leisure attractions which tended to spring up around cinemas.[42] In the latter half of the 1990s, many multiplex operators had begun once again to look at the urban centre as a potential site for the cinema. This renewed interest in city centres; a process called 'recentralisation' reflected a change in planning priorities for leisure-based developments between 1993 and 1997 under Prime Minister John Major's Conservative government.[43] In a series of Planning Policy Guidance notes (PPGs), which local planning authorities must consider in preparing their development plans, a new emphasis began to be placed upon regenerating urban centres. In 1994, *Transport Planning Policy Guidance Note 13 (PPG13)* sought to reduce private car use and promote the use of public transport by encouraging local authorities to "make maximum use of the most accessible sites, such as those in town centres."[44]

In June 1996, the new emphasis placed upon regenerating urban centres was explicitly signalled in *PPG6 Town Centres and Retail Developments*, which revised the first *PPG6* published in July 1993 and sought to roll back the trend for out-of-town developments, and whilst not outlawing them completely it sought to encourage edge-of-centre or town centre developments ('town centre' was used generally to cover city, town, and suburban district centres).[45] Moreover, *PPG6* was the first to make special mention of multiscreen cinemas, which suggested, according to Pal and Jones, that there were "special concerns about such developments."[46] Citing the case of two proposed multiplexes in Eastleigh in south Hampshire, which were refused planning permission in 1995, Pal and Jones suggested that "central government concerns that new multiplex cinemas should be located in town centres would seem to loom large."[47] *PPG6* instructed local authorities to adopt what it called a "sequential approach" to selecting sites for new development: "first preference should be for town centre sites, where suitable sites or buildings suitable for conversion are available, followed by edge of- centre sites, district and local centres and only then out-of-centre sites in locations that are accessible by a choice of means of transport."[48] A report by the Urban Task Force noted that "[t]oday, a national average of 70% of new development is on brownfield land, compared with 56% in 1997."[49] Brownfield sites referred to reclaimed industrial land.[50]

In 1997, the new Labour government under Prime Minister Tony Blair reinforced the commitment to developing urban centres and further discourage out-of-town developments. Under the new auspices of Deputy Prime Minister and Minister of Environment, Transport and the Regions, John Prescott, the stress was placed increasingly on that of 'sustainable development' with out-of-town developments increasingly seen as: encouraging excessive use of motorcars; having a poor public transport infrastructure; and building on valuable greenfield land. The decisive shift in policy came in 1999 with the publication of *A Better Quality of Life*.[51] In order to deliver sustainable development—defined as "ensuring a better quality of life for everyone, now and for generations to come"—planning regulations stressed the role leisure and shopping developments could have in sustaining the "vitality and viability" of urban centres, something *PPG6* had first outlined.[52] Leisure, including bars, restaurants, and multiplex cinemas, would encourage visitors in to the urban centre outside of normal shopping and work hours, with the explicit aim of creating a more vibrant night time economy.[53]

The changes in planning regulations that sought to stimulate the growth of cinemas and other leisure sites in urban centres also saw an increase in the number of residential developments in the city centre and an attempt to encourage, through planning and diversification of land use, an "urban renaissance."[54] According to the Urban Task Force, "[f]or the first time in 50 years there has been a measurable change of culture in favour of towns and cities, reflecting a nationwide commitment to the urban renaissance."[55] The perceived importance of encouraging residential development reflected a decades-long trend of depopulation in Britain's major city centres. According to the research and policy institute Centre for Cities, city centre populations grew by 37 per cent between 2001 and 2011, with 49 per cent of the residents of the large city centres (e.g. Manchester, Birmingham, Sheffield) aged between 20 and 29.[56] This large growth in the population of Britain's cities reflected their small size prior to the 2000s, with Manchester as a case in point since in 1990, 90 people lived in the heart of the city, which had grown to 25,000 residents in 2005.[57] Manchester became a major locus of multiplex development within the city centre during this period. UCI's Printworks development, located in a former newspaper printing works owned by The Daily Mirror Group, which had been closed by the company in the 1980s and lain derelict until the late 1990s, was transformed into a 300,000-square-foot leisure and entertainment centre in 2000. Manchester's City Council leader, Richard Leese, stressed the importance of the site saying that the centre would "broaden the interest and attraction of the city centre,"[58] whilst UCI's head Steve Knibbs felt that it would "be the nearest thing to Piccadilly Circus outside London."[59]

UCI already had a 20-screen, 4000-seat megaplex (for a while Britain's largest cinema) at the Trafford Centre, a large out-of-town shopping complex on the outskirts of Manchester, opened in 1998, and one of the country's top-grossing cinemas.[60] Nonetheless, UCI's Dave Harris felt that The Printworks would be the busiest in-town site in the country, drawing on a large population that included 100,000-plus students at the city's four universities.[61] The 110,000-square-foot cinema was designed by the Printworks' architects RTKL and was an eight-level structure that housed one of the largest IMAX screens in Europe.[62] The megaplex was also the first of a series of new sites rebranded as TheFilmWorks, which was followed by a 14-screen site in Greenwich in April 2001 and the refurbishment of CIC's old High Wycombe site. The new cinema concept was intended, according to UCI, to "deliver an experience beyond film" and incorporated larger screen, "black boxes" (auditoria with totally black décor to help focus attention on the screen), stadium seating and VIP balconies with "love seats," and a private bar.[63] In December 2001, AMC re-entered the British multiplex cinema market, when they opened a 16-screen megaplex—Great Northern 16—in a mixed-use development in a converted railway goods warehouse in the centre of Manchester. This meant that outside of London, Manchester had the largest concentration of screens, with 168 in 2002.[64]

Increasingly, the central government charged local authorities with developing plans for their town and city centres but though the rhetoric of sustainable development was new, the importance of commercial interests was still paramount. The development of new spaces for retail and leisure, as part of a broad strategy of "regeneration" in Britain's towns and cities, would rest ultimately on encouraging private developers in "partnership" with local authorities.[65] According to Crosby et al. between 1999 and 2003, eight large shopping centres opened in the middle of major towns and cities in Britain, including the Oracle Centre in Reading, a town located approximately 40 miles west of London; which contained department stores, shops, and leisure facilities including a ten-screen multiplex cinema.[66] Of the eight new complexes, The Chimes in Uxbridge near London incorporated a nine-screen multiplex, Touchwood in Solihull near Birmingham a nine-screen multiplex, and the Festival Place in Basingstoke a ten-screen multiplex. Cost consultants Davis Langdon & Everest were clear about the impact of changes of planning legislation on the siting of multiplexes: "[t]he effects of PPG6 and PPG13, planning policies aimed at controlling out-of-town development and reducing car usage, has been to direct opportunities for future cinema development back into urban locations."[67] In a report on the prospects for Britain's towns of the evening and late-night economy, The Civic Trust explicitly highlighted the role that the cinema could play:

> [c]inemas are some of the most inclusive of our evening economy spaces, appealing to teenagers, older people, young families and non-drinkers as much as to the population as a whole. Major new town and city-centre developments are reintroducing screens as part of the commercial mix... Westfield Shopping Towns and Land Securities see cinema and leisure very much as part of the commercial mix and have included multiplexes within their schemes for Swindon, Nottingham and Exeter.[68]

Many local authorities began to focus on what became commonly referred to as the "night-time economy," with the emphasis on entertainment and retail provision. Chief amongst its proponents was the think-tank Comedia and an influential research study of 15 town centres in which the group proposed the "eighteen-hour city" in which "the activities of the daytime are spread out into the evening."[69] *PPG6* set out the range of options for local authorities when considering planning policies and:

> encourage a range of complementary and evening and night-time economy uses which appeal to a wide range of age and social groups, ensuring that provision is made where appropriate for a range of leisure, cultural and tourism activities such as cinemas, theatres, restaurants, public houses, bars, nightclubs and cafes.[70]

By the early part of the 2000s, local authorities across Britain were engaged in concerted efforts to engender what Ellis-Reynolds identified as a "leisure-led renaissance" in their city centres.[71] Signalling what was clearly emerging as a shift of emphasis from out-of-town to city centres, AMC's head of European operations Bruno Frydman observed that:

> On the basis of my experience so far, I would say that the new factor which we have had to come to terms with in Europe is the need to work with town and city centres. This is a continent-wide mentality: in Italy and Germany, for example, the vitality of urban centres has always been a priority. But now the United Kingdom has also introduced legislation making multiplexes proposed for the suburbs of major cities subject to approval by the local authorities and requiring such schemes to meet criteria set by central government. Moreover, the Government now in power seems to be more sensitive to the issue of urban development than its predecessor.[72]

In 2002, *Screen Finance* reported that multiplex building had slowed down, in part as a result of a series of takeovers.[73] What was less clear from the article was that by 2000, the landscape for cinema development had changed, in part as a result of changes in the 1990s to the prevailing planning culture and a subsequent cultural and economic turn back towards the town and city centre as a site of retailing and leisure. Vanier quoted the British Film Institute's Paul Brett, who felt that there "was a lot of room for growth—especially in cities" and a "lot of space for building new sites."[74] In a paper given in 2001 at a conference on *Cinemas in the Community* Dodona Research's Karsten Grummitt informed delegates that for the first time the majority of new multiplex openings were in city centres or what were increasingly referred to as "out-of-centres" rather than out-of-town locations.[75] AMC's second multiplex opened in October 2003 in Birmingham, as part of its re-entry into the British market, was a case in point. With 12 screens, the Broadway Plaza complex was located at Five Ways in Edgbaston, in a converted children's hospital and adjoining site, on the city's inner ring road.

Ster Century, another new entrant to the multiplex market, also focussed on town centre sites, including the Castle Mall Shopping Centre in the centre of Norwich, opened in 2000, and The Light in Leeds, opened in 2002, which incorporated a 13-screen cinema and a range of restaurants, bars, and shops, as well as a health club, in what was the former head office building of the Leeds Permanent Building Society.[76] Ster Century was the European arm of the South African Ster-Kinekor company, which in 1999 was proposing to develop 44 sites in ten European countries. By 2004, it operated five multiplexes and one 16-screen mega-plex in Britain totalling 73 screens. It was purchased by rival Vue in May 2005, which re-branded all of the sites. In all, these cinema developments reflected a national trend as the number of screens in city centres or edge-of-centre cinemas increased by over 20 per cent between 2001 and 2004.[77] By 2017, the figure had increased by 26 per cent again over the 2004 figure, as 48.2 per cent of Britain's 4264 cinema screens were in town/city centres, 14.3 per cent in edge-of-centre sites, and 34.1 per cent in out-of-town sites.[78] That the majority of edge-of-centre cinemas are multiplexes is evidenced by the fact that the average number of screens per site there is 8.6 as opposed to 4.3 in town/city centre sites.[79]

For many of the large circuits the move to city and edge-of-centre sites, in particular, presented a range of challenges, not least the reduced space for construction. Unlike many leisure facilities that require large footprints, such as health and fitness clubs and ten-pin bowling, multiplex operators have been able to adapt their designs to suit smaller footprints by building multi-storey developments. Invariably, multiplexes have been seen almost exclusively as the most popular anchor tenant to mixed retail and leisure schemes.[80] The multiplex's importance to developers is that they help guarantee visitor numbers and offset the increased construction and devel-opment costs, which are approximately 20 per cent higher than compara-ble out-of-town sites.[81] The attractions of a mixed-use complex extend beyond the possible cost savings to encompass many of the perceived advantages of the out-of-town site, including adjacent car parking, shelter from the elements, proximity of other leisure attractions, and personal safety. Nonetheless, as Ennis-Reynolds argued, the shift from out-of-town to town-centre presented considerable risks to multiplex developers, including the:

- higher costs associated with the much higher design and fit-out spec-ification demanded by urban locations
- competition with other in-town leisure and retail uses for leisure expenditure compared with out of town, where the multiplex is usu-ally the anchor attraction
- need to recoup the greater costs quickly before obsolescence requires refurbishment; the large investment in development and the high rental values that result mean that any future use must also be of a high value[82]

Partly in response to these developments, several of the major circuits have launched up-market cinema brands, which have incorporated facilities such as licenced VIP lounges, reclining leather seats, and in-screen dining. National Amusements' Showcase Cinema de Lux brand was launched in the USA in 2008 with the opening of a 16-screen megaplex in Florence, Kentucky,[83] and by 2018 there were 14 Cinema de Lux sites in the USA. In Britain, Showcase opened its first such site—a 12-screen complex—in the Westfield Shopping Centre in Derby in May 2008. It was quickly followed by two more: a 12-screen site in Leicester's Highcross Shopping Centre and a 13-screen site in Bristol's Cabot Circus development, both opened later in 2008. All were marketed as more luxurious cinemas and had restaurants and licenced cinema screen called the Directors Lounge, with in-seat food and drinks service.[84] Showcase rather grandiosely talked of the 2000-seat Derby Cinema de Lux as "a concept set to redefine the U.K. cinema experience."[85] Significantly, these developments were all located at the centre of the three cities, as anchors within new shopping developments. In 2015, Showcase committed to spending some £10 million on retrofitting recliners in most of its sites as well as including them as standard in all new multiplexes.[86]

In 2017, Odeon, which is owned by AMC, began undertaking a series of conversions of multiplexes to what it called Odeon Luxe, which featured leather reclining seats, bars, new screens, 4K digital projection, and Dolby Atmos sound. The first conversion was the 12-screen East Kilbride site, built in 1996. When it opened it had 2559 screens with its biggest auditorium seating 428 and its smallest 89; however, after conversion, the larger seats and greater legroom meant that the total capacity was reduced to 941. Its largest auditorium now accommodated only 139, whilst its smallest a mere 35. This trend, imported from the USA, for larger seating and thus smaller capacities, was seen as a trade-off in terms of attracting audiences in urban areas who could increasingly go to a range of smaller "boutique" cinemas being opened by expanding circuits such as Curzon and Everyman. Audiences for the multiplex's luxury offer seemed prepared to pay more for the enhanced environment and the ability to pre-book a seat, which meant that although they had fewer seats these multiplexes had greater seat utilisation, meaning higher profitability.[87] This trend offers some sense of why Cineworld acquired the Picturehouse circuit in 2013 and why they retained that circuit's identity rather than fold it into the Cineworld brand (see Chap. 6).

Urban shopping and leisure developments like Highcross in Leicester were evidence of an observation Hopkins made, in his study of West Edmonton Mall in Canada, about the ways in which town centres were "moving indoors, both literally and metaphorically."[88] The 'lessons' of shopping mall developments in North America informed a delegation from Sheffield City Council, which was proposing to develop a large out-of-town shopping centre in July 1986, which subsequently became Crystal Peaks.[89] They visited a range of shopping malls in Canada, including the Eaton, Sherway, and Woodbine Centres in Toronto and the West Edmonton Mall. Ostensibly concerned with assessing the impact of out-of-town centres on cities and their downtown areas, the delegation noted with interest the value of adding leisure facilities in order to increase visitor attraction.[90] One of the key elements of the new Crystal Peaks site was the inclusion of the ten-screen AMC multiplex, opened in May 1988.[91]

Perhaps the most significant new multiplex circuit to emerge for many years, and one completely in tune with the new geography of cinemagoing, was The Light, formed by Tim Pullinger and John Sullivan in 2007. In the December of that year, they opened their first site—an eight-screen multiplex, as part of the Marine Point development in New Brighton on the Wirral peninsula. Marine Point was a mixed shopping and leisure scheme, which aimed to regenerate the town centre at what had been a popular seaside destination; however, more established circuits had passed on the opportunity to open there. Pullinger argued that with a population of 330,000 on the Wirral, the location was ideal for a smaller multiplex, whilst the installation of digital-only projection meant that the economics made sense.[92] The circuit has grown since 2007 by building new sites except for its nine-screen multiplex in Cambridge, which it purchased from Cineworld in 2015; so that by the end of 2018 it operated ten sites with 76 screens. It is due to open a further six multiplexes in 2019 with 49 screens and five in 2020 with 44 screens. Unlike its rivals, the company states that it "has and will, open in smaller towns, many of which have been previously unserved by multiplexes, A city/town centre location with easy access by both car and via safe pedestrian access to public transport."[93] According to Pullinger, their cinemas "give access to high quality entertainment in regional areas that don't necessarily have easy access to London and the West End," adding that "the Light is tempting back audiences that have been turned off by the generic multiplex experience offered by the market leaders."[94]

The impact of the *PPG6* planning guidance on multiplex building was acknowledged by Warner Village, when setting out its vision of three types of cinema in 1998: the "Giant Multiplex" as it described its 30-screen StarCity site (see Chap. 6); the "Multiplex" as typified by the stand-alone, 12-screen complex, located on an out-of-town site; and the "Citiplex," a compact building of six to ten screens located in the city centre, as typified by its Worcester site.[95] Opened in 1999, the site opposite the city's medieval cathedral was "bespoke" and designed to be "in keeping with the city's historic street patterns and local architecture."[96]

Given that the overwhelming majority of Britain's traditional cinemas are in town and city centres, the multiplex cinema is still located predominantly outside the urban centre. Nevertheless, the changes to planning laws in the past 20 years and a developing culture of "town centres first" have seen the re-emergence of the cinema as a key feature of the city and town centre. In part the attractions of the multiplex cinema in both out-of-town and centre rest with the relative attractions of both locations to cinemagoers, and these are determined by a range of demographic factors. In leisure and shopping developments in both locations, the multiplex cinema is seen as a major attraction and of universal appeal. According to research undertaken by Mintel Oxygen in 2010, "almost all types of consumers are more likely to have visited an out-of-town cinema than one in a city/town centre location, reflecting the greater prevalence of multiplex cinemas in out-of-town sites."[97] Nevertheless, there was also evidence of the re-emergence of cinema in the urban centre, since the same research suggested that cinemagoing attracted "more than half of all adults to town centres," signifying the perceived importance of cinemas to the developing urban leisure and retail economy.[98]

## Designing the Multiplex—"Not Just a Building, It's a Complete Concept"

When The Point opened in 1985, it was the first purpose-built, multiscreen cinema to open in Britain and the radical pyramid design was both a dramatic statement by the developer and a bold departure from previous cinema aesthetics. However, the Ziggurat structure was not the cinema, which was in fact a windowless, metal-clad, flat-roofed box, at the rear and linked by a glazed walkway from The Point in front. However, the site was dramatic, especially at night when the design emphasised the use of lighting: the "ephemeral construction materials of glass and electricity."[99] Nonetheless, the multiplexes built subsequently, with very few exceptions, were essentially big, rectangular metal and brick sheds, built on level sites.[100] Writing in 1999, Harriet Lane observed that though they had increased the opportunity for people to see films:

> the multiplex is a difficult creature to like … the giant Virgin, Warner Village, UCI and Showcase cinemas are charm-free cavernous constructions, often situated near motorway junctions or out-of-town shopping centres. Their architecture reflects their surroundings. Cruising past with your boot full of Ikea bags, you have probably mistaken them for storage depots or yet another furniture superstore.[101]

Indeed, whilst attending a multiplex opening in 1989 two distributors commented, anonymously, that they could have well been at an opening for a carpet warehouse rather than a cinema.[102] When asked about the prospects of a more adventurous architecture, Virgin Cinemas' Simon Burke argued in 1996 that cinemas were "best done as large square boxes—the acoustics and viewing angles are best if they're done that way, so we're not going to come up with spheres or wavy lines or anything kind of whacky."[103]

Initially, the designs for multiplexes built by the first round of US companies were essentially clones of the US plans, simply adapted to British standards. An example was that of AMC, which was taken over by UCI in 1988. Of particular concern was that, according to UCI managing director Ian Riches, many of the largest auditoria were too small, with seating capacities of less than the desired 500.[104] UCI, like Warner Bros. International Theaters (WBIT) and other circuits in the 1990s, began slowly to design for the British market but they did not always get it right. Though WBIT's in-house architect Ira Stiegler, who had designed shopping malls in the USA as well as a series of multiplexes before working exclusively for WBIT, argued that though largely similar to their designs in the USA, the standard of their multiplexes in Britain exceeded them. According to WBIT one of their multiplexes was "not just a building, it's a complete concept."[105] However, WBIT were still using sloped floors rather than the stadium seating that the company saw being used in Germany.[106] Having decided to incorporate stadium seating into their subsequent round of multiplex developments, the first in what would be a projected project pipeline was Leicester's nine-screen complex at Meridian

Park, designed by Stiegler and which was the first multiplex in the city, opened in July 1995. The cinema's internal layout also set the template for subsequent WBIT developments across Britain, with its "movie set" foyer, with references to Warner's Loony Tunes characters amongst others, and arrangement of nine screens, considered an optimum number in terms of operability. The nine screens, which incorporated the new stadium seating, were so arranged as to incorporate one projection room. Stiegler argued that the key to the building's design was the use of lighting designed to enhance what he referred to as "night architecture."[107] In 1989, Stiegler's 12-screen site in Bury,[108] near Manchester, had drawn the attention of several commentators, largely because of its use of glass and blue steel. Stiegler set out the design philosophy of the building as:

> essentially to fortify the sense of event and entertainment befitting a cinema [offering] the promise of an exciting space within by means of a glass lobby that telescopes out of the base of the building. Aglow at night, the lobby serves as a beacon of evening entertainment.[109]

Butler observed more acidly that at night, Bury's multiplex had "the squat authority of a cross-Channel ferry … with gold pyramid lights—like portholes—dotted around the exterior."[110]

Their position alongside shopping centres and on sites adjacent to motorways and major roads meant that multiplexes were viewed and approached from a distance. The iconography of the buildings was expressed in terms of logos, glitzy designs of steel, glass and coloured brick, and bright lighting. Aesthetically, the multiplex promised a place of entertainment that was in tune with its contents—namely the glamour and excitement of the feature film and, specifically, the Hollywood feature film.

For cinemagoers, the bright lights drew them in via the first point of contact, the foyer (or lobby in the USA). It was in the foyer, which assumed a large and prominent position within the overall structure, where the identity and ethos of the cinema were firmly established. Prominent were the box office and the concession counter, whilst the space served as a giant advertising and promotional zone for present and forthcoming films. Evidence suggested in 2002 that over 70 per cent of visitors to the multiplex spent up to 15 minutes in the foyer.[111] The foyer thus presented a unique challenge to designers: how to maximise sales of concessions and other goods, whilst at the same time moving several thousand people in and out of the auditoria and building.

Initially, the foyer had a separate box-office, located as the cinemagoer entered the building and a concession stand in the centre or towards the back. In addition, early designs of stand-alone sites might incorporate restaurants or bars off the foyer, such as AMC's initial designs. Following the experience of AMC at The Point, CIC, for instance, did not provide its own restaurants. Instead, it included space either side of the foyer, which was rented to a national pizza restaurant circuit and a company which opened a pub. This would be a model for all subsequent multiplexes that chose to incorporate these sorts of attractions. Indeed, manager Chris Green felt that the Wycombe 6 was quite unlike any other cinema at that time: "[i]n appearance, it is more like a modern shopping complex than a cinema."[112] The analogy that could be made between the multiplex and the shopping centre was intuitive since both sites determined spatially the relationship between consumer and provider, and the ways consumers perceived the site of exchange. For many people, the multiplex offered a similar environment to many others they had begun to spend time in, such as shopping centres, leisure centres, bowling alleys, and, of course, fast-food restaurants. Consumption as a driver for new forms of design was an expression of the ways in which neo-liberalism had begun to exert a powerful economic, social, political, and cultural influence in the 1980s and 1990s. Neo-liberalism determined not just the aesthetics of the new leisure spaces but also their location and operation.[113]

Nonetheless, much of the evidence suggested that the new multiplexes were popular with cinemagoers, with the CAA *CAVIAR 10* report revealing that the multiplex's "added extras," such as catering and bar/restaurant facilities were of interest to more than 50 per cent of patrons, whilst "comfortable cinema" was the third highest reason for visiting after proximity to home and choice of films.[114] Speaking in 1987, Douglas Hamilton, designer of the Maybox circuit's Slough multiplex, observed that "[w]hat was wrong with cinema product we couldn't do anything about. But, if we approached it from the point of view of the way people were treated, in terms of civility, hygiene, environment—if we approached it from the point of view of fun—then maybe there was something we could do about it."[115] Like many designers and architects, Hamilton's brief was to make the multiplex different from the cinemas that preceded them by focussing, in the words of David Fraser, of design company FITCH, on the "experience" their customers got from going to the cinema, adding that "instead of a product-based decision (going to a particular film) the consumer must be invited to make a lifestyle decision (going for an evening out)."[116] As the manager of Warner Village's O2 multiplex observed: "[t]here's been a revolution in the whole cinema experience. The cinemas of the past were just houses that showed the film. The multiplex offers a completely new dimension."[117]

Implicit in the history of cinema architecture, particularly from the 1930s onwards, have been the ways in which it reflected both modernist and post-modernist aesthetics, with the cinemas of Oscar Deutsch's Odeon circuit expressing the modernist principles of the age and The Point's Mesopotamian Ziggurat or StarCity's "fairground at night" examples of post-modernist pastiche. Moreover, the new multiplex designs involved some attempt to re-establish an identity for the cinema, in which the functional design which echoed the shopping centre was replaced by a more playful and post-modernist aesthetic which plagiarised many of the features of the older "picture palaces."

## The Multiplex and the Audience

The development of the multiplex had an immediate impact upon cinema admissions, as the CAA *CAVIAR 10* survey found that the percentage of people who go to the cinema had risen from 38 per cent in 1984 to 62 per cent in 1992, with half of regular cinemagoers visiting a multiplex cinema.[118] Moreover, the frequency of visits to multiplexes was greater than that of traditional cinemas.[119] In the CAA *CAVIAR 12* report, published ten years after 1984's nadir in cinema attendance, the influence of the multiplex was quantifiable since: "the proportion of visits to multiplex screen is significantly higher than one might expect from the proportion of screens that are multiplex."[120] Between 1996 and 1998, when multiplexes increased dramatically, the proportion of adults visiting the cinema increased from 58 to 72 per cent.[121]

The social profile of cinemagoers from the year multiplexes began opening in 1985 showed an increase in the number of ABC1s attending, whilst those classified as C2DEs decreased as a proportion of the increased audience size generally. The CAA *CAVIAR 10* report speculated on the reason for this trend, evident in 1993, citing: the impact of video and satellite television on "lower social grades"; the mix of films available; the kinds of cinemas that appealed to those with more disposable income; and the greater use of cars to travel to the cinema.[122] In 2017, the proportion of ABC1s attending the cinema was greater than the lower group across the cinemagoing population as a whole, though the proportions did not reflect the population as a whole. So, whilst 55 per cent of the population were classified as ABC1, amongst cinemago-ers they were over-represented at 62 per cent (amongst over-35s it was 67 per cent).[123]

On the eve of the multiplex in 1984, cinemagoing in Britain was domi-nated by younger adults as, according to Dodona (citing the CAA), 59 per cent of 15- to 24-year-olds 'ever went' (at least once) to the cinema whilst 15 per cent went at least once a month.[124] In the first decade after the multiplex arrived, the numbers of 16- to 24-year-olds going to the cinema grew, so that by 1993, for instance, 90 per cent of 16- to 24-year-olds were claiming to ever go, with 35 per cent claiming to go at least once a month.[125] Notwithstanding the risks of over-reporting in audience studies, in 1991 the British Film Institute (BFI) claimed that multiplexes were suc-cessfully appealing to those cinemagoers over 35 years old, who made up 27 per cent of the multiplex audience as opposed to 19 per cent at tradi-tional cinemas.[126] This reflected a more sustained increase in the propor-tion of over-35s going to the cinema, which has increased every year since the first multiplex opened, so that by 2017 it was just under 40 per cent.[127] There has been a small but steady decline in the number of younger adults (16- to 24-year-olds) going to the cinema, from 40 per cent in 1990 to 28 per cent in 2017, though they are still the largest cinemagoing demo-graphic group by far.[128]

From 1985, as the number of multiplexes increased so did the number of cinema admissions, which rose steadily overall, notwithstanding some peaks in certain years, which corresponded with certain blockbuster releases, to reach 177 million in 2018, the highest since 1970.[129] Britain now has an excess of multiplex cinemas. In 2017, there were 341 of them, from a total of 772 cinemas; however, multiplexes accounted for 78 per cent of screens.[130] Three companies dominate the multiplex market in Britain—Cineworld, Odeon, and Vue—which own 63.7 per cent of screens and take some 69 per cent of the annual total box-office, which was £1.28 billion in 2017.[131] Showcase, whilst a significant and enduring presence in the exhibition marketplace, have a much smaller market share—5.7 per cent from 6.5 per cent of screens—than any of their larger rivals.[132] These major multiplex operators—Cineworld, Odeon, Vue, and Showcase—then have been able to use their market dominance to maintain a strong pres-ence in all categories of sites, especially via their inclusion in new presti-gious city-centre developments across Britain, such as Showcase's Cinema de Lux complexes.

The emergence of the multiplex in Britain has revolutionised, perhaps modernised, the business of cinema management, with greater emphasis on customer service, comfort, and convenience. However, their business and employment practices have relied to a great degree on low-paid, de-skilled, and increasingly non-unionised staff. For Pearson, the arrival of the multiplex from the USA read like the plot of the 1950s science fiction film *The Day The Earth Stood Still*, with the omnipotent alien occupant, accompanied by a robot, replaced by an executive from a large exhibitor at a press conference who "demonstrates the company's intellectual and physical (read financial) power, warn what will happen to the British Film Industry if their company's plans are not put into operation, and departs, leaving a glossy brochure."[133] This articulation of the multiplex as the expression of Americanisation, in the pejorative sense, was much articulated, as we have seen, particularly by those representatives of the independent cinema sector. Nevertheless, it is also the case that the multiplex emerged in Britain at the very point at which cinemagoing was in terminal decline. As Moran posited, "multiplexes are a good thing," adding that they have "proved to be the saviour of cinemagoing in the UK, just as it did in the USA, a decade or so before."[134]

## NOTES

1. See Stuart Hanson, "A 'Glittering Landmark for a 21st Century Entertainment Centre': The Story of the Point Multiplex Cinema in Milton Keynes," *Historical Journal of Film, Radio and Television* 33, no. 2 (2013): 270–88.
2. "Half of All European Screens are Multiplexed," *Screen Digest* (November, 2003), 324.
3. "Global Cinema Exhibition Markets," *Screen Digest* (October, 2003), 301.
4. BFI, *BFI Statistical Yearbook 2018: Exhibition* (BFI Research and Statistics, 2018), 4. https://www.bfi.org.uk/sites/bfi.org.uk/files/downloads/bfi-screen-sector-certification-and-production-bfi-2018-09-03.pdf (accessed 22 November, 2018).
5. Urlan Wannop and Gordon E. Cherry, "The Development of Regional Planning in the United Kingdom," *Planning Perspectives* 9 (1994): 29–60, 39.
6. Susanne Cowan, "The People's Peace: The Myth of Wartime Unity and Public Consent for Town Planning," in *The Blitz and its Legacy: Wartime Destruction to Post-War Reconstruction*, eds. Mark Clapson and Peter J. Larkham (Farnham: Ashgate, 2013), 74.

7. See John Grinrod, *Concretopia: A Journey Around the Rebuilding of Postwar Britain* (London: Old Street Publishing, 2013).
8. Jim Yelling, "The Incidence of Slum Clearance in England and Wales, 1955–85," *Urban History* 27, no. 2 (2000): 234–54, 234.
9. Stuart Hanson, *From Silent Screen to Multi-screen: A History of Cinema Exhibition in Britain Since* 1896 (Manchester: Manchester University Press, 2007).
10. Ibid.
11. Peter Jones and David Hillier, "Changing the Balance—The 'Ins and Outs' of Retail Development," *Property Management* 18, no. 2 (2000): 114–26, 114.
12. Russell Schiller, "Retail Decentralisation—The Coming of the Third Wave," *The Planner* (July, 1986): 13–15.
13. Russell Schiller, "Vitality and Viability: Challenge to the Town Centre," *International Journal of Retail & Distribution Management* 22, no. 6 (1994): 46–50, 48.
14. Department of Transport, *National Travel Survey Statistical Release.* 26 July, 2018. https://www.gov.uk/government/uploads/system/uploads/attachment_data/file/729521/national-travel-survey-2017.pdf (accessed 24 November, 2018).
15. Department of Transport, *National Travel Survey Statistical Release.* 28 July, 2011, 10. http://assets.dft.gov.uk/statistics/releases/national-travel-survey-2010/nts2010-01.pdf (accessed 8 October 2011).
16. See Andrew Gamble, *Britain in Decline: Economic Policy, Political Strategy and the British State*, Fourth Edition (Basingstoke: Macmillan, 1994).
17. Neil Crosby et al., "A Message from the Oracle: The Land Use Impact of a Major In-town Shopping Centre on Local Retailing," *Journal of Property Research* 22, no. 2–3 (June–September 2005): 245–65.
18. Rod Griffiths, "Planning in Retreat? Town Planning and the Market in the Eighties," *Planning Practice and Research* 1, no. 1 (2010): 3–7, 5.
19. Clifford M. Guy, *The Retail Development Process: Location, Property and Planning* (London: Routledge, 1994), 70.
20. Ross Davies, "Retail Planning Policy," in *Cases in Retail Management*, ed. Peter McGoldrick (London: Pitman Publishing, 1994), 233.
21. Schiller, "Vitality and Viability".
22. By the time the Salford Quays multiplex opened in December 1986, TESE had sold its screen entertainment division and its cinemas had been purchased by the Cannon group in May for £175 million.
23. Hilaire Gomer, "Cannon Aims for Multi-screen Success," *The Guardian*, 24 July, 1985.

24. See Michelle Lowe, "Britain's Regional Shopping Centres: New Urban Forms?" *Urban Studies*, 37, no. 2 (2000): 261–74 and Michelle Lowe, "The Regional Shopping in the Inner City: A Study of Retail-led Urban Regeneration," *Urban Studies*, 42, no. 3 (March, 2005): 449–70.

25. In December 1988 AMC announced that they were to pull out of the UK market. They sold their eight multiplexes to a partnership of CIC and United Artists Communications, which was renamed United Cinemas International (UCI).

26. Edward Soja, *Thirdspace: Journeys to Los Angeles and Other Real-and-imagined Places* (Oxford: Blackwell, 1996), 238.

27. Barbara E. Phillips, *City Lights: Urban-Suburban Life in the Global Society*, Second Edition (New York: Oxford University Press, 1996), 150.

28. Joel Garreau, *Edge City: Life on the New Frontier* (New York: Anchor Books, 1992). Garreau proposed that one of the criteria for the 'edge city' was that it had, like all urban areas, "more jobs than bedrooms," meaning that the population increases at 9 am when people head towards it rather than away from it.

29. Paul Barker, "The Future Is Here and Now: Lots of Happy Smiling People Tripping to the Shopping Mall. But Does It Work?," *The Guardian*, 8 October, 1996.

30. Australian operator Hoyts opened a 12-screen multiplex in the complex in March 1999, the first in a planned expansion into Britain and Europe. However, the company opened no more and its British operation was deemed uneconomic. It sold the site to Showcase in 2000 which operates an expanded 17-screen, 2879-seat complex.

31. Ibid.

32. "The Point—Signalling a New Era in Cinema Exhibition?," *AIP & Co.* 78 (October, 1986): 29–31, 30.

33. Allen Eyles, "March of the Multiplexes," *Picture House* 14/16 (Spring, 1990): 22–25, 25.

34. See Stuart Hanson, "'Entering the Age of the Hypermarket Cinema': The First Five Years of the Multiplex in the UK," *Journal of British Cinema and Television* 14, no. 4 (2017): 485–503.

35. Ralf Ludemann, "Building a Dream Palace: From Site to Sound," *Screen International* 912 (18 June, 1993), 16.

36. Peter Jones and David Hillier, "Changing the Balance—The 'Ins and Outs' of Retail Development," *Property Management* 18, no. 2 (2000): 114–26, 117.

37. Alan Collins, Chris Hand, and Andrew Ryder, "The Lure of the Multiplex? The Interplay of Time, Distance and Cinema Attendance," *Environment and Planning* 37 (2005): 483–501.

38. CAA, *Cinema and Video Industry Audience Research (CAVIAR) 20.* (London: CAA, 2003).

39. Tina McFarling, "Cinema Exhibition Discussed at BFI Conference Seminar," *Screen International* 516 (28 September, 1985), 11.

40. Dodona, *Cinemagoing 11* (Leicester: Dodona Research, 2003), 16.

41. Martin Wroe, "Pre-packed Fun in the Pleasure Dome," *The Observer*, 8 October, 1995.

42. Bruno Frydman, "Exporting the Multiplex Model to Europe: The Experience of AMC" at the MEDIA Salles Round Table. *The Impact of Multiplexes on the Cinema Market and on their Environment*, Amsterdam: Cinema Expo International, 15 June, 1998. http://www.mediasalles.it/expo98fr.htm (accessed 10 September, 2011).

43. Neil Ravenscroft, "The Vitality and Viability of Town Centres," *Urban Studies*, 37, no. 13 (December, 2000): 2533–49, 2533.

44. Department of the Environment, *Transport Planning Policy Guidance, PPG13* (London: HMSO, 1994), para. 21.

45. Department of the Environment, *Town Centres and Retail Development, Planning Policy Guidance, PPG6* (London: HMSO, 1996).

46. John Pal and Peter Jones, "Multiplexes—What's the Picture?" *Town and Country Planning*, 65, no. 12 (December, 1996): 344–5, 345.

47. Ibid., 345.

48. Department of the Environment, *Town Centres and Retail Development*, para. 1.11.

49. Urban Task Force, *Towards a Strong Urban Renaissance*. An independent report by members of the Urban Task Force chaired by Lord Rogers, November, 2005, 2. http://www.urbantaskforce.org/UTF_final_report.pdf (accessed 7 October, 2011).

50. See Stuart Hanson, "From Out-of-Town to the Edge and Back to the Centre: Multiplexes in Britain from the 1990s," in *Watching Films: New Perspectives on Movie-Going, Exhibition and Reception*, eds. Albert Moran and Karina Aveyard (Bristol: Intellect, 2013).

51. Department of Environment, Transport and the Regions, *A Better Quality of Life—A Strategy for Sustainable Development*, Cm 4345 (London: DETR, 1999).

52. Ibid., para. 1.1.

53. See Georgina Ennis-Reynolds, "Sustainable Development and Multiplexes," *Journal of Leisure Property* 2, no. 4 (2002): 317–31.

54. Office of the Deputy Prime Minister, *Planning for Town Centres, Planning Policy Guidance, PPG6* (London HMSO, 2005).

55. Urban Task Force, *Towards a Strong Urban Renaissance*, 2.

56. Elli Thomas, Ilona Serwicka, and Paul Swinney, *Urban Demographics: Where People Live and Work* (London: Centre for Cities, July, 2015), 1.

57. Urban Task Force, *Towards a Strong Urban Renaissance*, 1.

58. Quoted in John. J. Parkinson-Bailey, *Manchester: An Architectural History* (Manchester University Press, 2000), 263.
59. Mary Scott, "Intercity Express," *Screen International* 1131 (24 October, 1997), 12.
60. Mary Scott, "StarCity Highlights Megaplex Concerns," *Screen International* 1269 (28 July, 2000), 2.
61. Mary Scott, "Executive Suite: Dave Harris," *Screen International* 1273 (25 August, 2000), 12.
62. With the purchase of Odeon by AMC in 2016, the company came to an agreement with the Competition and Markets Authority (CMA) to sell the site to rival Vue. The CMA threatened "an in-depth investigation unless AMC could address concerns over the merger's effect on the competition between cinemas in Manchester." See Lee-Anthony Bennett, "Manchester's Odeon Cinema in the Printworks is closing down," *Viva*. http://vivamanchester.co.uk/manchesters-odeon-cinema-in-the-printworks-is-closing-down/ (accessed 24 January, 2019).
63. "The Film Works for UCI Cinemas," *Designweek*, 1 September, 2000. https://www.designweek.co.uk/issues/31-august-2000/the-film-works-for-uci-cinemas/ (accessed 20 November, 2018).
64. Dodona, *Cinemagoing 11*, 24.
65. See Peter Jones, David Hillier, and Daphne Comfort, "Urban Regeneration Companies and City Centres," *Management Research News* 26, no. 1 (2003): 54–63.
66. Crosby et al. "A Message from the Oracle," 246.
67. Davis Langdon & Everest, "Cost Model," *Building*, 5 May, 2000. http://www.davislangdon.com/upload/StaticFiles/EME%20Publications/CostModels/MultiplexCinemas_CM_5May00.pdf (accessed 12 October, 2011).
68. The Civic Trust, *Night Vision: Town Centres for All* (London: Civic Trust, October, 2006), 60.
69. Comedia, Out of Hours: A Study of Economic, Social and Cultural Life in Twelve Town Centres in the UK (London: Comedia, 1991), 22.
70. Office of the Deputy Prime Minister, *Planning for Town Centres*, para. 2.23.
71. Ennis-Reynolds, "Sustainable Development and Multiplexes," 327.
72. Frydman, "Exporting the Multiplex Model to Europe."
73. Fiona Vanier, "Multiplex Numbers Fall for First Time in 2002," *Screen Finance* 16, no. 6 (26 March, 2002): 4–5.
74. Ibid., 4.
75. Cited in Ennis-Reynolds, "Sustainable Development and Multiplexes," 318.

76. Ster Century was the European arm of the South African Ster-Kinekor company, which in 1999 was proposing to develop 44 sites in ten European countries. By 2004, it operated five multiplexes and one 16-screen megaplex in the UK totalling 73 screens. It was purchased by rival Vue in May 2005.

77. UK Film Council, *Statistical Yearbook/Annual Review 2004/05* (London: UK Film Council, 2004), 40.

78. BFI, *BFI Statistical Yearbook 2018: Exhibition*, 16.

79. Ibid., 16.

80. See Mintel Oxygen, *In- vs Out-of-Town Leisure* (Mintel International Group, November, 2009).

81. Davis Langdon & Everest, "Cost Model."

82. Ennis-Reynolds, "Sustainable Development and Multiplexes," 324–5.

83. This cinema was sold to rival Rave Cinemas in 2009 and purchased by Cinemark in 2013.

84. See Hanson, "From Out-of-Town to the Edge and Back to the Centre."

85. Andreas Fuchs, "Star Entrance," *Film Journal International* 112, no. 2 (February, 2009): 14–17, 15.

86. Patrick Von Sychowski, "Every Multiplex Is New Again," *Cinema Technology* 31, no. 1 (March, 2018), 22.

87. Dodona, *Cinema Industry Research UK & Ireland* (Leicester: Dodona Research, 2018), 6.

88. Jeffrey. S. P. Hopkins, "West Edmonton Mall as a Centre for Social Interaction," *The Canadian Geographer* 35, no. 3 (September, 1991): 268–279 cited in Ravenscroft, "The Vitality and Viability of Town Centres," 2535.

89. See Hanson, "Entering the Age of the Hypermarket Cinema."

90. Sheffield City Council, *Report of Director of Land and Planning to Planning and Transportation Programme Committee* (7 July, 1986).

91. See Hanson, "Entering the Age of the Hypermarket Cinema."

92. "Seeing the Light," The Completely Group, 22 April, 2015. https://completelygroup.com/article/seeing-the-light (accessed 10 December, 2018).

93. The Light, http://www.lightcinemas.com/whatwedo/complete-night-out (accessed 13 December, 2018).

94. Rebecca Burn-Callander, "Indie Cinema Chain to Challenge UK's 'Homogenous' Giants," *The Daily Telegraph*, 30 June, 2015.

95. "10 years of Warner Bros. International Theatres: Promotional Feature," *Screen International* 1150 (20 March, 1998), 28, 30.

96. Ibid., 30.

97. Mintel Oxygen, *In- vs Out-of-Town Leisure*.

98. Ibid.
99. Nataša Ďurovičová, "Los Toquis, or Urban Babel," in *Global Cities: Cinema, Architecture, and Urbanism in a Digital Age*, eds. Linda Kraus and Patrice Petro (New Jersey: Rutgers University Press, 2003), 80.
100. Hanson, *From Silent Screen to Multi-screen*.
101. Harriet Lane, "Fancy a Film Tonight?: Does the Out-of-Town Multiplex Offer More than Just Pricey Popcorn and Nachos?," *The Observer*, 24 January, 1999.
102. Tina McFarling, "The Plex Factor," *Producer* 12 (Summer, 1990): 21
103. Quoted in Leslie Felperin, "Multiplexity," *The PACT Magazine* 56 (September, 1996), 17.
104. Quoted in Ralf Ludemann, "Building a Dream Palace: Sitting Pretty," *Screen International* 912 (18 June, 1993), 28.
105. *Warner Bros. Cinema Corporate Profile Brochure*, in Andreas Fuchs, "Entertaining Designs: Architect Ira Stiegler Capitalizes on WB Brand," *Film Journal International* 101, no. 12 (1 December, 1998), 106.
106. Quoted in Fuchs, "Entertaining Designs," 104.
107. Philip Turner, *Warner Cinemas: An Outline History* (St. Paul's Cray: Brantwood Books, 1997), 19.
108. This was closed in 2010 and demolished in 2016.
109. Quoted in Kevin Lally, "A Smashing British Debut for WB Int'l: Manchester 12-Plex Heralds Ambitious Programme," *The Film Journal* 92, no. 9 (1 October, 1989), 32.
110. Robert Butler, "A Night at the Pictures in 1994," *Independent on Sunday*, 16 October, 1994.
111. CAA, *CAVIAR 20*.
112. Quoted in Tina McFarling, "Wycombe 6," *Screen International* 610 (25 July, 1987), 16.
113. See Hanson, *From Silent Screen to Multi-screen* and "Entering the Age of the Hypermarket Cinema."
114. CAA, *Cinema and Video Industry Audience Research (CAVIAR) 10* (London: CAA, 1993), 8, 28.
115. "Hamilton Approaches Slough with 'Fun'," *Screen International* 628 (28 November, 1987), 23.
116. Quoted in Felperin, "Multiplexity," 17.
117. Lee Parry, manager of Warner Village Multiplex in the O2 centre in London, quoted in Lane, "Fancy a Film Tonight?"
118. CAA, *Cinema and Video Industry Audience Research (CAVIAR) 10*, (CAA: London, 1993), 7–8.
119. Ibid., 24.
120. CAA, *Cinema and Video Industry Audience Research (CAVIAR) 12*, (CAA: London, 1995), 21.

121. Jamie Doward, "Multiplexes are Cinemas' Paradiso," *The Observer*, 27 September, 1998.
122. CAA, *CAVIAR 10*, 21–22.
123. BFI, *BFI Statistical Yearbook 2018: Audiences* (BFI Research and Statistics, 2018), 6. https://www.bfi.org.uk/sites/bfi.org.uk/files/downloads/bfi-statistical-yearbook-audiences-2018-12-17.pdf (accessed 22 November, 2018).
124. Dodona, *Cinemagoing 4*, 6.
125. Ibid., 6.
126. BFI, *Film and Television Handbook 1992* (London: British Film Institute, 1991), 39.
127. BFI, *BFI Statistical Yearbook 2018: Audiences*, 5.
128. Ibid., 5.
129. UK Cinema Association, "2018 Admissions Figures Confirm an Extraordinary Year for UK Cinemas," 22 January 2019. https://www.cinemauk.org.uk/2019/01/2018-admissions-figures-confirm-an-extraordinary-year-for-uk-cinemas/ (accessed 27 January, 2019).
130. Dodona, *Cinema Industry Research: UK & Ireland 2018* (Leicester: Dodona Research, 2018), 7.
131. BFI, *BFI Statistical Yearbook 2018: Exhibition*, 22–23.
132. Ibid., 22–23.
133. Nichola Pearson, "Multiplexing: They came from America," *New Socialist* (October/November, 1988), 53.
134. Mark Moran, "Multiplexes Are a Good Thing," *Cinemabusiness* 20 (December, 2005), 13.

# PART III

# The Multiplex Cinema in Europe

# CHAPTER 8

# The Multiplex in Belgium, the Netherlands, and Scandinavia

Having opened the first multiplex in Britain, US exhibitors also looked to mainland Europe to expand, in part because of the penetration and popularity of mainstream, Hollywood films, allied to the sophisticated organisation of US distributors. Interviewed in 1992, Joost Bert, head of Belgian exhibitor Kinepolis, and Dieter Buchwald, an independent cinema owner from Germany, observed that the exhibition sector in Europe was "totally dependent on US products" with US distributors supplying "85–93 per cent" of films.[1] In Europe, circuit building on the scale of the USA had not taken place, with *Screen Digest* reporting in 1991 that in the USA there were nine circuits with more than 500 screens but only one in Europe, and that was the fragile MGM/Pathé/Nordisk/Cannon consortium with 536 screens.[2] One of the obstacles to building large circuits in Europe, apart from the obvious issues of working across national boundaries, was the difficulty in securing a steady supply of films. The exhibitors from the USA were looking for new markets, in the knowledge that they could command an adequate supply of films from these globalised, US distributors; many of whom were part of vertically integrated conglomerates like Warner Bros.

There seemed good reason to suppose that Europe was, in the words of United Artists' vice president Mal Birnbaum, a "fertile marketplace," not least given what seemed to US exhibitors to be the perilous state of cinema exhibition in many Western European countries.[3] As Hollywood looked to increase its foreign box-office there was a feeling that Western Europe

© The Author(s) 2019
S. Hanson, *Screening the World*,
https://doi.org/10.1007/978-3-030-18995-2_8

was under-screened, with about one third of the number of screens per capita as the USA even though it had about the same population. Indeed, one American exhibitor observed that "the situation overseas is ripe for a revolution. The theaters in Europe are like the theaters in the States used to be 40 years ago: mostly terrible. But give people a reason to go to the movies more often and they will ... especially to American movies."[4] In Western Europe, the number of screens had been consistently falling throughout the 1970s and early 1980s, from 35,035 in 1970 to 23,396 in 1985, whilst in the USA the same period saw a growth in screens from 13,750 to 21,147.[5]

By 1995, the largest exhibitors in Western Europe were United Cinemas International (UCI) (joint-owned by Universal and Paramount) and Warner Bros. International Theaters (WBIT) with operations in five countries each—Austria, Britain, Germany, Ireland, and Spain in the case of UCI, and Denmark, Germany, Portugal, Spain, and Britain in the case of WBIT. No existing European circuit operated in more than one country in addition to its domestic market. Nevertheless, the growth in multiplex screens in Western Europe doubled in the five years prior to 1995, whilst the number of screens not in multiplexes increased by just 4 per cent.[6]

In a survey by London Economics for MEDIA Salles on the obstacles to expansion into Europe, the views of a group of British-based cinema companies were sought.[7] The survey indicated that the:

> reason the American chains had targeted the UK for multiplex development is the common language, and relative similarity in business cultures. Progress elsewhere in Europe was seen as fraught with the difficulties both of language and of breaking into the existing tightly knit exhibition sectors.[8]

Having considered Britain, where new multiplex construction had doubled admissions within the first ten years, US companies nonetheless moved more easily into some continental European markets such as Germany, the Netherlands, and Denmark, but found entry into others considerably more difficult, especially Belgium, Norway, and France. This is not to suggest that foreign exhibitors could not access these markets, since countries like Belgium, for instance, had a significant presence from French exhibitors. However, the companies interviewed by London Economics identified a set of obstacles in some northern countries in Europe:

Holland, Germany and France, have firm policies regarding development of their cities. Authorities are opposed to the building of multiplexes on the outskirts of cities, on the grounds that this would attract people to the suburbs, thereby increasing the ongoing economic deterioration of the city centres. This is both an economic and a social policy.[9]

It is issues like these that offer a set of particular considerations which illustrate the development of the multiplex, and the role played by US and overseas companies. This chapter considers developments in Belgium, the Netherlands, and the countries within Scandinavia, especially Sweden, Norway, and Denmark. Indeed, though American Multi-Cinema's (AMC) The Point was the first overseas multiplex built by a US exhibitor, it was not the first purpose-built multiplex in Europe. By the time it opened in December 1985, SF Bio had opened four multiplexes in Sweden, whilst by 1981, the giant Decascoop cinema, in Ghent, Belgium, had opened its doors.

## BELGIUM

In 1950, cinema admissions stood at 116.36 million, declining largely year on year to a nadir of 15.66 million in 1988. Since then the total has fluctuated somewhat from the high of 25.3 million in 1998 to the current total in 2017 of 21.0 million.[10] In 1950, there were 1415 screens in Belgium, which peaked at 1585 in 1957 before declining to 412 in 1987 and reaching 496 in 2017.[11] The county is divided into three regions—the French-speaking Wallonia and the Flemish-speaking Flanders and Brabant (greater Bruxelles)—with Flanders, which contains the major cities of Antwerp, Ghent, and Bruges, containing 48 per cent of the country's screens and accounting for 46 per cent of all cinema visits.[12]

Though The Point was the first US-owned multiplex in Europe, Belgium can lay claim to opening Europe's first multiplex and the world's first megaplex cinema. Belgium has one of the most mature cinema markets in Europe in terms of multiplex development, due in large part to the Kinepolis Group under its owner Albert Bert, which introduced the multiplex concept in Europe. Like AMC's Stanley Durwood, Bert was a key figure in the development of multiscreen cinema, not just in Belgium. The Bert family's involvement in cinema stretched back to 1927 when Albert's grandfather, Charles, opened the Hôtel de Flandre cinema in Harelbeke and his father, Albert Snr, opened the Theatre Majestiek in the same town in 1941. Bert took over the running of the Majestiek in the 1968 and in 1970 he converted it into a twin: Belgium's first multiscreen cinema. Having partnered with his sister-in-law Rose Claeys in 1972, they opened the three-screen Trioscoop in Hasselt; to which they added four more screens subsequently and in 1975, they opened the five-screen Pentescoop in Kortrijk, which Bert argued was the first significant assault on what were characterised as the "shoeboxing sins" of previous cinemas.[13] In 1981, the company opened the 12-screen, 3800-seat Decascoop in Ghent, which Bert observed as the "transfer point from a mid-size complex to a multiplex."[14] The auditoria ranged from 90 to 580 seats with a single, computerised projection room, which could show one print in multiple auditoria. As Van de Vijver observed the complex was visited by several representatives of US exhibitors, who were keen to look at its architectural design.[15]

In 1987, Bert founded the Decatron company and a year later the company's dominance of the exhibition sector in Belgium was cemented with two developments: the merging of the Bert and Claeys companies to form the Kinepolis Group and the construction of the 24-screen, 7606-seat Kinepolis in Brussels. When it opened, the Kinepolis was the world's largest cinema and the first dedicated megaplex. In its first 12 months it sold more than two million tickets and in 1989 accounted for 52 per cent of the cinema audience in Brussels.[16] The Brussels megaplex heralded what Bert called the "Kinepolis style"[17] as the company opened an even larger complex in Antwerp—the 24-screen, 8690-seat Metropolis, in 1993. Kinepolis was adopted as the brand name for all of the multiscreen cinemas in the Bert and Claeys groups in 1995, whilst their formula for the Kinepolis Group rested, according to Lotze, on "three pillars" which were "economisation of the business operation, quality and mobility of customer."[18] "Economisation" referred to the economies of scale necessary to operate a multiplex including design, labour costs, and flexible programming. "Quality" was predicated on innovation and a consideration of the whole experience of visiting the cinema; whilst "mobility" was the extent to which the multiplex afforded ease of access and provided large amounts of free parking.[19]

One of Kinepolis' largest competitors in Belgium in the 1980s was George Heylen's Rex Concern circuit, which operated mainly in the Antwerp area, but also in Ghent and Bruges.[20] Unlike Kinepolis' focus on edge-of-town and out-of-town developments, adjacent to shopping developments, Heylen concentrated on town centre cinemas, and thus he and

Bert rarely competed head to head. In Antwerp in 1983, for instance, Heylen owned 27 cinemas with "10,000 cinemas seats within a radius of 250 yards."[21] Though Heylen was one of the first exhibitors to operate multiscreen cinemas they were largely conversions of older cinemas, with auditoria that sometimes had as few as 100 seats. Heylen's focus on small, city centre cinemas in the face of the steady development of the multiplex, along with the deterioration of both his cinemas and many of the areas in which they were located, led to the demise of his circuit. As Lotze explains, Kinepolis' first direct attempt to enter the Antwerp cinema market with a large megaplex had begun in 1989 with an initial planning application to the local authorities.[22] When that cinema—the Metropolis—opened in October 1993, it was shortly after Heylen had declared bankruptcy and his remaining 17 cinemas were closed. The Metropolis was initially ten screens but was shortly thereafter expanded with the addition of 14 screens, seating some 9000 cinemagoers in total.

Heylen had been resolutely a regional player; however, French company UGC (Union Générale Cinématographique), which entered the Belgian marketplace in 1978 with the opening of its eight-screen City II in Brussels, became Kinepolis' biggest competitor. Indeed, up until 1988 and opening of the Kinepolis megaplex, UGC held a wholly dominant position in the capital, controlling 50 out of 60 screens in 1984.[23] By 2000, Kinepolis and UGC controlled 46 and 42 per cent of the Brussels market, respectively, and, whilst UGC expanded beyond the capital opening multiplexes and acquiring others via takeovers, it was second to Kinepolis in terms of sites and screens after 1988. It had three sites and 43 screens to Kinepolis' ten sites and 121 screens in 2000,[24] whilst, by 2018, UGC had consolidated its position as number two with the purchase of rival Utopia and control of 72 screens as opposed to Kinepolis' 138.[25] The Belgian market, according to Biltereyst and Van de Vijver, has a significant level of concentration in the hands of these two players.[26] Though a relatively small market nationally, the Kinepolis Group assumed a dominant position, largely via its construction of multiplexes and megaplexes on the outskirts of major cities, so that within ten years of opening their Brussels megaplex, their 120 screens were more than their next four largest competitors combined.[27] Just a year later, in 1999, Kinepolis' share of the Belgian cinema market passed 50 per cent, though the company had attracted the interest of Belgian competition authorities who placed a 20 per cent cap on future investment in any one site.[28]

For Neckebroeck, Kinepolis constituted a "quasi-monopoly" in which the technical and business innovations, which had contributed to the modernisation of the Belgian exhibition industry, were countered by the tendency for cinemas to be located on the outskirts of major conurbations and a narrowing of the films available—largely mainstream Hollywood product.[29] Subsequent expansion of Kinepolis has been based upon entry into foreign markets, notably France and Spain, but also Italy, the Netherlands, and Switzerland. In France, it is the fifth largest exhibitor with 128 screens and operates the country's largest multiplex, a 23-screen complex in Lomme, Lille.[30] The company operates two giant megaplexes in Spain—the 25-screen, 9200-seat Kinepolis Madrid, opened in 1998 and acknowledged as the largest cinema in the world; and the 24-screen, 8200-seat Kinepolis Valencia. In 2017, Kinepolis moved beyond Europe with the purchase of Canadian exhibitor Landmark; Canada's second largest circuit with 44 sites and 303 screens, in a deal worth C$122.7.[31]

## THE NETHERLANDS

Like all of its neighbours, the Netherlands experienced a peak in cinema admissions in the 1950s with 64.2 million visits in 1958, followed by decline until the 1970s, when the figure stabilised somewhat, before a further decline from the beginning of the 1980s until 1992, when just 13.6 million people went to the cinema.[32] After this low point admissions recovered, though it was not until the 2000s that they began to rise more dramatically so that, for instance, between 1999 and 2003 they went up more than 33 per cent, albeit from a very low base.[33] In 2017, admissions reached 36 million, which was the highest since the mid-1980s and equated to 2.1 visits per year per capita.[34] As in all countries in Europe, the number of screens in the Netherlands has declined since the 1960s from a peak of 565 in 1961 down to 415 in 1993.[35] With approximately 400 screens for 15 million people, the Netherlands was considered to be the most under-screened of the major European territories.[36] As in Britain, the Netherlands' first multiplex cinema—a six-screen, 1684-seat complex built by MGM International Cinemas on Wilhelminastraat in Maastricht and opened in September 1994—appeared at a time when cinemagoing was in a poor state. Like Britain, the company that opened it was an overseas one. MGM International Cinemas was the European arm of Giancarlo Parretti's Pathé Communications Corporation, which had purchased Cannon cinemas, the largest circuit in the Netherlands in 1987, and then purchased MGM/United Artists in 1990. The Cannon circuit had been re-branded MGM by 1991, with MGM International Cinemas as the collective name of its European operations.[37]

Cannon had been the Netherlands' dominant circuit immediately prior to the opening of the first multiplex, with 34 per cent of the exhibition market in 1988 but 95 per cent of major cities such as Amsterdam, Rotterdam, and The Hague.[38] It had entered the market itself through the acquisition of the Tuschinski circuit's 11 cinemas in 1984. This meant that when MGM opened the multiplex in Maastricht, it was the dominant force in exhibition, with 20 sites and 69 screens. MGM also set out a series of multiplexes planned for opening within the next few years, including a six-screen site in Groningen and a nine-screen multiplex in Amsterdam, adjacent to the historic Theatre Tuschinski.[39] However, within 12 months the MGM brand had disappeared. With the collapse of Parretti's company and the passing of ownership to the Crédit Lyonnais bank, MGM International Cinemas was auctioned in 1994 and acquired by French company Chargeurs, owners of Pathé (see Chap. 9). When the proposed MGM multiplex opened in Groningen in November 1995, it was a Pathé and had nine screens.

Overnight, Pathé became the largest circuit in the Netherlands by some considerable margin. Prior to its acquisition of MGM in 1993, Pathé's owner, Chargeurs, had entered into a race to open the Netherlands' first multiplexes by allying with US companies Morgan Creek and Warner Bros, International Theaters (WBIT). The venture, called Warner-Morgan Creek-Chargeurs, outlined proposals to invest $20 million to build three multiplexes in Amsterdam, Rotterdam, and beginning with an eight-screen complex called MovieWorld in Scheveningen, close to The Hague, opened in April 1995. Whilst acknowledging that MGM had opened their Maastricht site several months earlier, the company claimed that MovieWorld was the country's first "state-of-the-art multiplex."[40] Millard Ochs, WBIT's President, argued that the market potential in the Netherlands was untapped and, with reference to his previous role with AMC and their entry into the European market via Britain; he argued that "the revolution in Europe had started," with MovieWorld and cinemas like it set to "become a social entertainment destination point for visitors."[41]

Upon its purchase of MGM, Pathé effectively dissolved its deal with WBIT and Morgan Creek, and assumed control of the MovieWorld site, adding it to its rebranded sites and set about replacing many of its older cinemas with multiplexes. These included the Pathé ArenA in Amsterdam, which with 14 screens and 3250 seats was the biggest in the Netherlands when it opened in April 2000 and remains so. Pathé also grew by acquiring other exhibitors such as rival Minerva Bioscopen in 2010. Pathé's domination of exhibition in the Netherlands was built upon its multiplex-building programme. By 2004, ten years after the first multiplex opened, eight of the county's sixteen multiplexes were operated by Pathé, including near-monopolies in Amsterdam and Rotterdam.[42] By 2017, Pathé owned 25 sites with 204 screens, 13 per cent and 23 per cent, respectively, but took 44.2 per cent of the total box-office.[43]

Pathé's domination of the multiplex sector and control of exhibition in the major cities meant that its competitors relied more on multiscreening their existing sites, with the addition of some small-scale multiplex building. At the time when MGM were building their multiplex in Maastricht, the second largest exhibitor was Jogchem Theaters BV (also known as JT Bioscopen), which operated 11 cinemas with 46 screens.[44] Throughout the 1990s and into the 2000s, Jogchem Theaters did not invest in new multiplexes, only beginning to build in 2011, when they opened some of the Netherlands' most technically advanced multiplexes, in Vlaardingen, Kerkrade, Hoorn, Eindhoven, and Hilversum. In 2010, Jogchem had become the first circuit to commit to converting its entire circuit to Digital Cinema Initiative (DCI)-compliant digital projection systems, in a deal with the integrator XDC. The seven-screen Media Park complex in Hilversum, opened in October 2014, was the first to instal Barco's 3D, laser-illuminated projectors, whilst its eight-screen Eindhoven multiplex, opened in December 2015, was one of the first in the world to instal the Dolby Cinema system.[45] In August 2015, having embarked on its multiplex building programme, Jogchem was sold to British-based exhibitor Vue, for €85 million, which acquired 21 cinemas and 111 screens across 20 towns and cities; making it the second largest exhibitor in Europe after Odeon/UCI.[46] The acquisition was Vue's fourth in three years, having purchased Britain's Apollo in 2012; Germany and Denmark's CinemaxX, also in 2012; Poland's Multikino in 2013; and Italy's largest cinema circuit Space Entertainment in November 2014.[47] As the second largest cinema circuit in the Netherlands, after Pathé, Vue controlled 21 sites with 118 screens in 2017 and accounted for 12.8 per cent of the Netherlands' box-office.[48]

The Netherlands' third largest exhibitor is the Belgian company Kinepolis, which acquired the Wolff Bioscopen circuit's nine sites and 46 screens, plus two construction projects for multiplexes in Dordrecht and Utrecht, in August 2014.[49] Kinepolis had previously entered the Netherlands market via short-lived agreements with domestic exhibitors, having a 35 per cent share, along with local company Cinema Groep, in the seven-screen, CineCity multiplex opened in Vlissingen in 1999, and a 51 per cent share in the five-screen CineMec complex, opened in Ede, in 2000

Kinepolis opened the Wolff-planned Dordrecht site, a six-screen, 1200-seat multiplex, in February 2016 and the 14-screen, 3200-seat Kinepolis Jaarbeurs complex in Utrecht in May 2017 (having opened the first six screens in December 2016). This was almost the same size as the Pathé ArenA site in Amsterdam, though it was considerably more technologically advanced, with laser projection in all auditoria. In between these projects, inherited in part from Wolff, Kinepolis opened the ten-screen, 1727-seat Breda multiplex. Its position as the Netherlands' third largest exhibitor was entrenched with the acquisition of the Luxemburg-based circuit Utopia in August 2016, and its five Dutch sites. Kinepolis opened a seven-screen complex in Hertogenbosch in June 2018, which meant it had 16 sites with 111 screens. In 2017 (prior to the opening of its latest site), the company accounted for 9.6 per cent of the box-office.[50] Though smaller than Vue, its average of seven screens per site, as opposed to Vue's 5.6, suggests Kinepolis' cinemas are larger.

The adoption of the multiplex in the Netherlands was slower than in many countries in Europe for a variety of reasons. In the first instance, domestic exhibitors were reluctant to invest in new building, preferring to divide up their existing cinemas into multiscreen sites. In part, this was based upon the low rates of cinema attendance in the country, though the slow rates of multiplex opening were reflected in a parallel growth in admissions, which increased consistently from the early 2000s as the multiplex began to take root. Indeed, according to Dodona "historically the Dutch have never been among Europe's high cinema-attending nations but this is starting to look like the consequence of conservative investment in the past by the country's exhibitors rather than a matter of preference on the part of the public."[51] If one considers the rollout of multiplexes in the Netherlands in the 1990s, this is evidenced by the increases in local attendance in town and cities when a local multiplex was opened. In Maastricht, Den Haag, Groningen, Rotterdam, and Eindhoven, for example, admissions per person, according to Dodona, rose after the entry of the multiplex.[52]

That said, the slow rates of multiplex building, particularly in the first ten years after the first multiplexes opened in 1994, were also to do with the regulatory environment in the Netherlands and in particular the lengthy planning processes. Many of the Netherlands' largest cities, such as Amsterdam and Utrecht, did not see any multiplex developments until some years after the first sites were opened. In Britain and Germany, multiplex construction in the early days took place largely out of town or on the edge of towns. In the Netherlands, developers who preferred sites away from the centre ran up against many city councils which were reluctant to approve such developments, preferring that exhibitors looked for sites nearer the centre. However, this inevitably presented another problem, which was the potential impact upon existing cinemas. Many councils, such as Amsterdam's, preferred exhibitors to renovate their existing downtown cinemas, rather than build on the edge of the city, which also had an impact on the diffusion of the multiplex. However, finding town and city centre sites was difficult though, with several exhibitors expressing frustration at the extent to which, in the words of Joachim Wolff, President of the Wolff cinema circuit, "cinema building in Holland is a very political thing. It's best to keep a low profile until you've cleared the hurdles."[53]

In Vlissingen, Kinepolis, along with local exhibitor Cinema Groep, spent two years planning the seven-screen Cine City multiplex, as, according to Morrison, "bureaucratic stumbling blocks delayed construction."[54] Endeavouring to enter the Netherlands market earlier, Kinepolis had proposed a 16-screen megaplex in the Amsterdam suburb of Dieman at the same time as rival Pathé was planning its 14-screen ArenA multiplex in the south of the city (see above). Diemen city council gave permission for the project but it was turned down by the Province of North Holland, on the basis that two large multiplexes would have too dramatic an impact upon the city. This precipitated a lengthy legal battle to secure planning permission, which was never forthcoming.

The exhibition sector in the Netherlands reflects some significant trends, which have included the wide diffusion and entrenchment of the multiplex, so that 52.7 per cent of screens are in multiplexes, which number 27 in total and account for 46.5 per cent of cinema visits in 2017.[55] With 48 per cent of the screens and 66.6 per cent of the box-office, exhibition in the Netherlands is now also highly concentrated in the hands of three foreign-based companies—French-owned Pathé, Belgian-owned Kinepolis, and British-owned Vue.

## SCANDINAVIA

In 1998, two exhibitors—Sweden's Sandrews and Denmark's Metronome, owned by the Norwegian Schibsted group—merged to form The Sandrew Metronome Corporation, placing a notice in the trade press which read "from now on you can consider the Scandinavian countries one market."[56] The interrelationships between exhibitors in Denmark, Norway, and Sweden, largely driven by mergers and acquisitions in the last 20–30 years, has meant that the penetration of multiplexes has been substantial, but unlike Germany and Britain, for instance, the direct entry of US exhibitors has been virtually nil.

Sweden can lay claim as the location of the first purpose-built multiplex in Europe, when in February 1980 Svensk Filmindustri opened the Filmstaden (Film City) in Stockholm. The cinema, located in a shopping centre near the TK department store, had 11 screens and could seat 1148 people, with the two largest auditoria accommodating 173 people and the smallest 61. When it opened, the Filmstaden was not the largest multi-screen cinema in Europe though, since in the Danish capital Copenhagen, the Palads Bio, operated by Nordisk Film and originally opened in 1918, had been subdivided into 12 screens seating a total of 1800 in 1978. In 1979, a further five screens were added in the building's basement, making the total seating capacity 2230. In Norway, the first multiscreen cinema was a conversion of an existing cinema—Oslo's Saga Theatre—which was rebuilt with six screens, the largest of which accommodated 598 people. Perhaps the forerunner of the purpose-built multiplex in Norway was the four-screen Edda Kino in Haugesund opened as a single-screen cinema in 1979 but by 1981 had two additional auditoria constructed, which was increased to four in 1984.

## DENMARK

In Denmark, as in all European countries, the trend for cinemagoing throughout the post-war period was down, from a peak in 1953 of 58.9 million admissions (12.1 visits per head) to just 8.65 million in 1992 (1.69 visits per head).[57] Since this low point, admissions rallied somewhat, but also fluctuated, so that by 2009 they had reached 14 million, the highest since 1982, before falling off to 11.9 million in 2017.[58] The drop in audience was matched by a contraction in the Danish cinema infrastructure, though whilst the screen count fluctuated after the high of 1959 (465 screens), the number in 1989 (318) suggested that when one took into account the dramatic fall in the number of seats over the same period (159,237 to 59,000) the evidence suggested widespread dividing up of previously single-screen sites.[59] From 1989, the number of screens continued to decline until 1999 when thereafter it began to rise, so that by 2017 there were 458 screens in 166 cinemas.[60] Of these cinemas, 25 were multiplexes of six screens or more.[61] As a small country with only four cities with populations over 100,000 people, Denmark's exhibition sector has historically been centred upon its capital city Copenhagen, which in 1994 on the eve of the multiplex accounted for some 40 per cent of admissions.[62] Copenhagen continued to dominate exhibition, though the growth of multiplex building in other cities, especially Odense for instance, meant that the proportion has dropped consistently, to 32 per cent of all admissions in 2002 and 24 per cent in 2017.[63]

Historically, the exhibition market in Denmark was controlled by two vertically integrated companies: Nordisk and Metronome. Largest by some margin was Nordisk, which in 1987 operated the 17-screen Palads multiscreen cinema in Copenhagen, in addition to eight other cinemas in the capital. By the late 1980s, Nordisk accounted for 75–82 per cent of the country's exhibition market,[64] and in 1989, it sold 50 per cent of the company to Giancarlo Parretti's Pathé Communications Corporation to form Pathé-Nordisk. When Parretti purchased MGM/UA in 1990, the exhibition subsidiary in Denmark was renamed MGM-Nordisk, with the company owning eight sites with 44 screens, though the MGM brand was not used on its cinemas.[65] Nordisk merged with the Danish media and publishing group Egmont in 1992 and with the collapse of Parretti's empire—and the auction of MGM's various exhibition subsidiaries—Nordisk purchased MGM's former 50 per cent holding in 1996 to regain control of the whole company. In the same year, Nordisk opened Denmark's first purpose-built multiplex—The Metropol—an eight-screen site in the centre of Odense (Denmark's third largest city). Additionally, it began a major renovation of its Palads site in central Copenhagen and subdivided further sites. In 2002, Egmont merged Nordisk and its other entertainment division, Egmont Entertainment, into a new, vertically integrated company called Nordisk Film. At this point, Nordisk Film was still the largest cinema circuit in Denmark, with 43 per cent of the market.[66] The new company had just begun to build a series of new BioCity-branded multiplexes, with its first the six-screen BioCity in Hillerod in 2000, followed by the ten-screen BioCity in Aalborg in 2003.

The opening of Nordisk's multiplex in Odense precipitated a wave of multiplex construction with other exhibitors entering the market; indeed, in the same city distributor All Right Films opened the seven-screen Rosengrad Bio in December 1998. The company was a short-lived exhibitor but it did sell its Odense multiplex to the German circuit CinemaxX, which entered the Danish market in 2000. CinemaxX opened the first purpose-built multiplex with eight or more screens—CinemaxX Fisketorvet—on the site of a former fish market in central Copenhagen, in October 2000. This was followed by an eight-screen complex in Arhus, opened in October 2003. CinemaxX were significant because as a new entrant in the market, they were the only exhibitor in Denmark to operate a circuit of completely purpose-built multiplex sites. They were the largest exhibitor in Germany, though a series of financial crises in the mid-2000s, and eventual acquisition by British-based exhibitor Vue in 2012, meant that though they did not retreat from the Danish market, they did not expand their circuit either.[67]

Nordisk's only other significant competitor had been the Anglo-Swedish Sandrew Metronome, which in 2000 operated two cinemas in Denmark; the five-screen Dagmar Teatret, specialising largely in arthouse films, and the seven-screen Scala. The latter was opened in March 1989 as a five-screen cinema (enlarged with two additional screens in 1992) on the top floor of a new shopping centre opposite the entrance to the Tivoli Gardens. This was Denmark's first purpose-built multiscreen cinema and the first new cinema built in the city for some time. Like MGM's involvement in Nordisk, Sandrew Metronome had opened the Scala and operated the Dagmar in a 50:50 partnership with Warner Bros. International Theaters (WBIT). Indeed, the Scala was WBIT's first cinema opened in Europe. Sandrew Metronome's first, purpose-built multiplex—the 11-screen Kinopalaeet—was opened in the Copenhagen suburb of Lyngby in 2001, though with its buyout of WBIT's share of the business in 1999, the Kinopalaeet opened under its own brand. In 2005, owner Schibsted decided to sell its exhibition interests in Sandrew Metronome, which were purchased by Nordisk in May 2006. This left Nordisk in a totally dominant position in the Danish exhibition market, with 28.5 per cent of screens and 45 per cent of admissions.[68] In addition to its position within the Danish market, it was one of the first exhibitors to expand into the liberalising Norwegian market when it acquired Oslo Kinematografer's cinemas, in April 2013 (see below). This meant that by 2018, Nordisk operated 20 cinemas with 152 screens in Denmark.

## NORWAY

Though the exhibition landscape in Scandinavia is dominated by pan-regional companies, the circumstances that prevail with regard to individual countries have some interesting differences. In Norway, the country in which the purpose-built multiplex, especially by foreign owners, was slowest to emerge, the dominance of the municipal cinema was a considerable factor. Though, as Asbjørnsen and Solum point out, the ownership of cinemas by local authorities, counties, or municipalities was not unique to Norway, it was the "*extent* of the municipal cinema business, which since the late 1920s has held around 90% of Norwegian cinemas' turnover (in 1997 it was 93%)."[69] The system arose out of legislation passed in 1913 and with the National Association of Municipal Cinemas (NAMC) (Kommunale Kinematografers Landsforbund) established in 1917 (renamed Film & Kino in 1998). It was not until the 1960s and a fall in cinema revenues was predicated on the arrival of television that the system began to be questioned, though largely in relation to the competing demands of public service television. Cinema was given exemption from competition legislation on the grounds of cultural policy.[70] Nonetheless, given the nature of Norway's disparate population of some four million and the perceived necessity of subsidising cinemas that lay outside of the major areas of population density, the principle was extremely resilient.

From the 1970s, the municipal system came under more sustained pressure from the forces of marketisation and privatisation, at least in the media and communications sector. In 1981, in an echo of developments in Britain, a new Conservative Government in Norway signalled its shift to greater marketisation by deregulating the broadcasting sector and in the 1990s private cinemas began to appear, in many cases with the sale of municipal cinemas.[71] Moreover, in several cases municipalities turned their exhibition outfits into limited companies, which allowed the larger Scandinavian exhibitors to purchase shares in them, though in the case of the largest and most important—the state-owned Oslo Kinematografer, which had 30 per cent of the national market—the city did not allow such sales and sought a Norwegian partner.[72] These developments marked a watershed, with a number of pan-Scandinavian companies looking to enter the Norwegian market via partnerships. Nevertheless, in December 1997 Norway's first privately owned, purpose-built multiplex—the eight-screen Kino 1 opened in Sandvika, near Oslo—was built by the Norway Cinema Group, an investment company owned by Robert Johnsen.

Johnsen also operated a four-screen cinema in Arendal and saw the Kino 1 as a first step in an expanding circuit of multiplexes. Within two years of opening the Kino 1, with half-yearly admissions in 2000 of 198,487, was the fifth most successful cinema in Norway in terms of both box-office and admissions.[73]

As the new millennium dawned, 90 per cent of cinemas were still owned by municipalities, with Johnsen's Kino 1 group the leading private circuit. In 2000, Swedish-based SF Bio (the exhibition arm of Svensk Filmindustri) became the first foreign exhibitor to enter the Norwegian cinema market when it acquired a six-screen cinema in Tonsberg and, though it sought licences to build in Oslo, Tromso, Bergen, and Lillestrom, amongst others, it was not until 2006 that it opened seven- and five-screen multiplexes in Lillestrom and Moss respectively. SF Bio's president and CEO Jan Bernhardsson observed in 1999 that "we are seeing chinks in the municipal system, and we aim to be there to help fill the gaps."[74] This foothold in Norway was increased, when in 2008 SF Bio purchased a substantial part of Johnsen's Kino 1 group, which meant that by the end of the year it was the largest exhibitor in Norway. The tipping point for the substantial privatisation of Norway's cinemas came in April 2013 when the city council approved the sale of Oslo Kinematografer's cinemas to the Danish vertically integrated company Nordisk Film, owned by Egmont (see above). In 2014, the recently renamed SF Kino, then part of a Swedish/Finnish holding company called the Nordic Cinema Group, acquired 49 per cent of the municipal cinemas in Stavanger, Bergen, and Sandnes. This meant that these two companies then controlled 130 of Norway's 422 screens and accounted for 52 per cent of admissions.[75]

As part of the Nordic Cinema Group, SF Kino was subsequently acquired by US company AMC (itself owned by Dalian Wanda) in January 2017, meaning that the US exhibitor has become the largest exhibitor in Norway, in terms of screens but the second largest, after Nordisk, in terms of admissions.[76] As part of AMC, the Nordic Cinema Group became a subsidiary of the ODEON Cinema Group and from March 2018 changing its name to ODEON Kino. It opened its first such branded multiplex—the 14-screen ODEON Oslo in Storo on the edge of the city centre—in March 2018. The multiplex, the largest cinema in Norway, incorporated an IMAX screen, which was also the first in the Nordic countries to incorporate laser projection.[77] The company began to then rebrand all of its SF Kino cinemas similarly. The ODEON Oslo was also significant since it marked the first competition for rival Nordisk in the Oslo, which had a monopoly in the city. It has no comparable cinema to the ODEON 14-screen multiplex, though it operates the purpose-built, six-screen Ringen Kino on Sannergata.

## SWEDEN

In Sweden, the development of the multiplex, as we have seen, was considerably more mature, with its exhibition landscape highly concentrated and dominated by a small number of major players. Moreover, in contrast to developments in Britain and Germany, for instance, multiplexes have largely been built in the centres of cities rather than on the edge. Having opened Sweden's first multiplex in 1980, Svensk Filmindustri, which was purchased by the Bonnier Media Group in 1983 (with the exhibition arm renamed SF Bio in 1987), was the largest exhibitor by far. By the end of the 1980s, SF Bio operated 60 cinemas, including 19 multiscreen sites, and controlled 65–70 per cent of the Swedish market, with SF Bio's Mats Kullander attributing much of their success to the "Filmstaden concept."[78] The 11-screen Stockholm prototype had been, according to Kullander, "a simple multiplex, with several small auditoria instead of one large one" which was "geared towards the youth audience, with the Hollywood Inn restaurant, video and record stores, a trendy hairdresser and a fashion boutique," what he characterised as a "film supermarket."[79] Two of the 140-seat auditoria were operated by the Swedish Film Institute, with the complex attracting some 700,000 people per year, which resulted in a further four screens being added in 1984, taking the total seating to 1850.[80] SF Bio opened a number of eight-screen multiplexes in Vesteras, Helsinborg, Sundsvall, Norrköping, and Linköping, whilst also beginning a major improvement scheme for a further 64 cinemas. SF Bio maintained its dominance of the Swedish exhibition sector throughout the 1990s, opening ever-larger multiplexes under its Filmstaden brand, including a further 12-screen multiplex in Hotorget, Stockholm, and a 12-screen multiplex in Uppsala. SF Bio was one of the largest exhibitors in Scandinavia, with a significant presence in Norway (see above) and in March 2013 it was merged with the Finnish company Finnkino, owned by a Swedish private equity company called Ratos, to form the Nordic Cinema Group. This created a significant regional player as Finnkino dominated not just the Finnish exhibition sector but those of Estonia, Latvia, and Lithuania as well. Upon its formation, the Nordic Cinema Group controlled a total of 603 screens and sold over 27.1 million tickets, whilst in

Sweden SF Bio accounted for 80 per cent of admissions.[81] In 2015, SF Bio opened the Filmstaden Scandinavia, a 15-screen megaplex in the Mall of Scandinavia, in Stockholm, which included the Nordic region's first IMAX screen.

SF Bio's only significant competitor in terms of multiplex construction was Sandrews, which had been operating cinemas since 1926 and was also a vertically integrated film company. In the 1970s it had made a virtue of converting many of its cinemas into triple and quadruple screens, and in doing so claimed to have opened Scandinavia's first multiscreen cinema—Stockholm's three-screen Biografen Grand—in 1970.[82] Though Sandrews was a significant second place to SF Bio, its focus on building new multiplex cinemas in major cities such as Gothenburg, Malmö, and Stockholm in particular meant that it took a considerable market share in these areas. In 1991, it opened the ten-screen, 1148-seat BioPalatset multiplex in Stockholm (the first in Sweden to incorporate THX sound) and immediately accounted for 40 per cent of the capital's market.[83] This was followed by a six-screen multiplex in Malmö in 1994 and a ten-screen complex in Gothenburg in 1995. With the opening of these multiplexes, Sandrews, with a national share of 19 per cent of admissions, took 34–40 per cent in these three cities.[84]

The intensity of the competition between Sandrews and SF Bio, especially in the multiplex sector in the major cities, saw Sandrews file a legal complaint with the Swedish competition commission accusing SF Bio parent company Bonnier of blockading Sandrews' access to major Hollywood films for its newer cinemas, via its distribution company Svensk Filmindustri.[85] As vertically integrated companies, both Sandrews and SF Bio distributed films, though with 8–10 per cent and 35 per cent of the market, respectively, SF Bio had a significant advantage. This situation changed in 1998 with the merger of Sandrews with Danish exhibitor Metronome, itself owned by the Norwegian media group Schibsted, creating a large regional exhibition, distribution, and video company operating in Norway, Denmark, Sweden, and Finland. The exhibition arm was renamed Sandrew Metronome and included multiplexes in Copenhagen and Helsinki, whilst the company also distributed films for Warner Bros. across Scandinavia.

In 2005, Sandrew Metronome's parent company decided to concentrate on film distribution and put its cinemas up for sale. SF Bio sought to acquire the company's Swedish cinemas but the deal, which would have given SF Bio 95 per cent of the country's screens and control of all of the country's multiplexes, was blocked by competition authorities. Instead, its 22 Swedish cinemas were acquired by Malmö-based Triangelfilm who branded the cinemas as Astoria. Despite the best efforts of the authorities to engender competition in the Swedish exhibition market, financial problems just a year later saw Triangelfilm sell its cinemas outside of Stockholm to BF Bio with an additional 20 screens acquired a year later in 2007.

Unlike other territories in Europe—notably Germany and Britain—US multiplex operators did not directly enter the Scandinavian market, considering it marginal in terms of return, though they did form partnerships with Danish exhibitors (see below). The only operator to build a multiplex was AMC, which opened what was the largest complex in Scandinavia, the 18-screen, 4166-seat Kungens Kurva complex in the Heron City development, south of Stockholm, in 2001. AMC's megaplex was a key element, along with the world's biggest IKEA store, of a 36,000-square-metre shopping and entertainment centre, built by British property developer Gerald Ronson. The development of Kungens Kurva, in the municipality of Huddinge, was an initiative on the part of the local authority that sought to stimulate the local economy and bring employment, and was coincidentally near the headquarters of SF Bio. AMC's megaplex was the first cinema built in Sweden outside of the urban area and aroused much industry attention. When, in 1999, AMC had publicly announced that its proposed new cinema would be 24 screens, SF Bio's Stefan Klockby said that he was:

> not overjoyed about the new plex being built less than 300 meters from our back door, but he added the exhibitor would most certainly influence moviegoing in Stockholm if it ever got built. We have seen a lot of these projects on the drawing board but they don't always end up breaking ground.[86]

Though the eventual cinema was smaller, AMC confidently predicted that admissions would top one million within the first year; however, in practice the complex admitted just under three quarters of that number in its first full year of operation.[87] Tickets sold at weekends equated to a capacity of 80 per cent but on weekdays this figure dropped to 20 per cent, in part, it was argued, because the major distributor in Sweden, Svensk Filmindustri, part of the same group as SG Bio, refused AMC access to its own productions and co-productions.[88] For AMC, the disappointing admission figures meant that in December 2003, the company decided to exit the Scandinavian market and sell the site to SF Bio, who rebranded it Filmstaden Heron City. This gave them an unassailable dominance in the Stockholm area, since they had just opened Filmstaden Vista, an 11-screen multiplex in the north of the city, to complement the 14-screen Filmstaden Sergel in the city centre. In an effort to compete with SF Bio, Nordisk, which dominated the Norwegian and Danish exhibition markets, entered the Swedish market in December 2018, with a new five-screen multiplex in Gränbystaden, near Uppsala. Nordisk also has a proposal for another five-screen site in Mobilia near in Malmö, to be opened in 2020.

Despite some early efforts to enter the Scandinavian market via partnerships or via building as in the case of AMC in Sweden and Warner Bros. in Denmark, no US exhibitors were able to sustain their position in the rapidly conglomerating market. AMC's failure to maintain its position and grow especially seemed, at that point, to have signalled an end to US efforts to break into the Scandinavian market. However, this would not be the case since although no US companies did build there, AMC Theatres, itself owned by the Chinese Dalian Wanda group (see Chap. 4), purchased the Nordic Cinema Group in January 2017. AMC acquired 68 cinemas, with 463 screens and approximately 68,000 seats, in addition to a minority investment in another 50 associated cinemas with 201 screens.[89] Though AMC announced that it would maintain a headquarters in Stockholm, the company would operate as a subsidiary of the pan-European ODEON Cinemas Group, based in London.[90] This has meant that when one considers Nordisk's unassailable position within the Danish market and its significant presence in the Norwegian market, the region (including Finland and the Baltic states) is dominated by these two companies: AMC and Nordisk.

## NOTES

1. Kimberley Strub, "The European Film and Exhibition Climate," *BoxOffice* 128, no. 6 (1 June, 1992), 16.
2. "Cinema in Europe: Circuit Building and Multiplexes," *Screen Digest* (July, 1991), 153.
3. John Hazelton, "Theatrical performances," *Screen International* 617 (12 September, 1987), 16.
4. Adam Dawtrey "The Resurrection of British Cinema," *The Hollywood Reporter International Exhibition Special Report* 317, no. 25 (May, 1991), S-4.

5. "Cinemas and Their Audiences: Just Holding On," *Screen Digest* (September, 1992), 202.
6. "Market Profile: European Multiplex Cinemas—The Second Revolution," *Film Journal International* 99, no. 6 (1 July, 1996), 75.
7. London Economics/BIPE, *White Book of the European Exhibition Industry: Vol 2* (1994). http://www.mediasalles.it/whiteboo/wb2_1_3.htm (accessed 4 January, 2019).
8. Ibid.
9. Ibid.
10. Dodona, *Cinemagoing Western Europe 2000* (Leicester: Dodona Research, 2000), 56 and Dodona, *Cinema Industry Research: France & Benelux 2018* (Leicester: Dodona Research, 2018), 1.
11. Dodona, *Cinemagoing Western Europe 2000*, 50 and Dodona, *Cinema Industry Research: France & Benelux 2018*, 1.
12. Statbel, Exploitatie van bioscoopzalen in België, 2015–16. https://statbel.fgov.be/en/themes/enterprises/cinemas#figures (accessed 25 August, 2018).
13. Andreas Fuchs, "A Multi-screen Pioneer," *Film Journal International* 100, no. 6 (1 July, 1997), 106.
14. Ibid., 106.
15. Lies Van de Vijver, "The Cinema is Dead, Long Live the Cinema!: Understanding the Social Experience of Cinema-Going Today," *Participations: Journal of Audience & Reception Studies* 14, no. 1 (2017): 129–44.
16. Christian de. Schutter, "Think Big," *Screen International* 734 (9 December 1989), 34.
17. Fuchs, "A Multi-screen Pioneer," 110.
18. Kathleen Lotze, "Bringing the Multiplex to Antwerp: A Battle of Two Giants," *Tijdschrift voor Mediageschiedenis* 21, no. 1 (2018), 76–101, 80. http://www.tmgonline.nl/index.php/tmg/article/view/339 (accessed 30 July, 2018).
19. Ibid.
20. According to Lotze (ibid., 99) the characterisation of Heylen's company as the Rex Concern is incorrect since in the first instance it refers to the fact that his flagship cinema was called the Rex and secondly that the complex structure of Heylen's ownership meant that "Heylen's cinema group was not a concern, as the different companies that fell under Heylen's supervision could not technically be considered a 'business unit'."
21. Ted Clark, "Heylen Circuit Renovates Sites and Expands To 60 Screens," *Variety* 311, no. 1 (4 May, 1983), 451.
22. Lotze, "Bringing the Multiplex to Antwerp."
23. "Benelux-Films: Power Concentrations Increased Among Belgian Cinema Owners," *Variety* 316, no. 12 (17 October, 1984), 58, 62.

24. Andy Stern, "UGC, Kinepolis Duke It Out for Belgian Dominance," *Variety* 379, no. 5 (19 June, 2000), 56.
25. Dodona, *Cinema Industry Research: France & Benelux* 2018, 2.
26. Daniel Biltereyst and Lies Van de Vijver, "Cinema in the 'Fog City'. Film Exhibition and Social Geography," in *Cinema Beyond the City: Small-town and Rural Film Culture in Europe*, eds. Judith Thissen and Clemens Zimmermann (London: British Film Institute, 2016).
27. Marlene Edmunds, "Kinepolis Keeps the Plexes Coming," *Variety* 371, no. 6 (15 June, 1998), 74.
28. Andy Stern, "Kinepolis Still Rules the Belgian Scene," *Variety* 375, no. 6 (21 June, 1999), 64.
29. Kjell Neckebroeck, "France and Benelux: The Exhibition Basics," *The Film Journal* 99, no. 1 (1 January, 1996), 32.
30. Dodona, *Cinema Industry Research: France & Benelux 2018*, 13.
31. Kinepolis, *Kinepolis Completes Acquisition of Canadian Movie Theatre Group 'Landmark Cinemas'*, Kinepolis Group Press Release. https://corporate.kinepolis.com/en/press-releases/kinepolis-completes-acquisition-canadian-movie-theatre-group-landmark-cinemas (accessed 25 August, 2018).
32. Michel Gyory and Gabriele Glas, *Statistics of the Film Industry in Europe* (Brussels: The European Centre for Research and Information on Film and Television (CERICA), 1992), 205; and Edward Borsboom, *100 Years of Cinema Exhibition in the Netherlands—A Historical Profile*. Media Salles. http://www.mediasalles.it/ybkcent/ybk95_nl.htm (accessed 3 January, 2019).
33. Dodona, *Cinemagoing Benelux 2004* (Leicester: Dodona, 2004), 67.
34. Nederlands Vereniging Van Bioscopen En Filmtheaters. https://www.denvbf.nl/english/ (accessed 3 January, 2019).
35. Gyory and Glas, *Statistics of the Film Industry in Europe*, 204; and "World Cinema: Upward Reversal of Fortune," *Screen Digest* (September, 1994), 202.
36. "Three Pact to Build Multiplexes in the Netherlands," *Variety* 351, no. 2 (10 May, 1993), 93.
37. See Philip Turner, *Cannon Cinemas: An Outline History* (St. Paul's Cray: Brantwood Books, 1997) and *MGM Cinemas: An Outline History* (St. Paul's Cray: Brantwood Books, 1998).
38. Bruce Alderman, "Cannon In Holland has 68 Screens & 34% Market Share," *Variety* 332, no. 11 (5 October, 1988), 88.
39. Mary Van De Kerkof-Flanagan, "Dutch Theater Boom Continues," *Variety* 356, no. 13 (24 October, 1994), 66.
40. Pat Kramer, "Dutch Treat," *BoxOffice* 131, no. 7 (1 July, 1995), 42.
41. Ibid., 42.

42. Dodona, *Cinemagoing Benelux 2004*, 61.
43. Ibid., 20.
44. Thijs Faas, "Building an Industry," *Screen International* 863 (26 June, 1992), 14.
45. Dolby Laboratories, *JT Eindhoven Unveils One of the First Dolby Cinema Locations in the World* (3 December, 2014). http://investor.dolby.com/news-releases/news-release-details/jt-eindhoven-unveils-one-first-dolby-cinema-locations-world (accessed 3 January, 2019).
46. Vue International, *Vue International Acquires Second Largest Cinema Chain in the Netherlands* (21 August, 2015). http://vue-international.com/index.php/our-media/press-releases/84-vue-international-acquires-second-largest-cinema-chain-in-the-netherlands (accessed 3 January, 2019).
47. Tom Grater, "Vue Cinemas Acquires Dutch Exhibitor JT Bioscopen," *Screen Daily* (24 August 2015). https://www.screendaily.com/news/vue-cinemas-acquires-dutch-exhibitor-jt-bioscopen/5091967.article (accessed 3 January, 2019).
48. Dodona, *Cinemagoing Benelux 2004*, 20.
49. Kinepolis, *Kinepolis Takes Over the Wolff Cinema Group in the Netherlands* (22 July, 2014). https://corporate.kinepolis.com/en/press-releases/kinepolis-takes-over-wolff-cinema-group-netherlands (accessed 4 January, 2019).
50. Dodona, *Cinemagoing Benelux 2004*, 20.
51. Dodona, *Cinema Industry Research: France & Benelux* 2018, 17.
52. Dodona, *Cinemagoing Western Europe 2000*, 152–3.
53. Quoted in Marlene Edmunds, "Pathe plans Dutch plexes," *Variety* 375, no. 6 (21 June, 1999), 63.
54. Melissa Morrison, "Rise of the Netherlands: Vlissingen venture Opens a New Era in Dutch Exhibition," *BoxOffice* 135, no. 4 (1 April, 1999), 155.
55. Nederlands Vereniging Van Bioscopen En Filmtheaters, *Jaarverslag 2017* (Amsterdam: NVBF, 2017), 35.
56. "Step into the Future with Us," *Variety* 371, no. 6 (15 June, 1998), 63.
57. Gyory and Glas, *Statistics of the Film Industry in Europe*, 55 and "World Cinema: Upward Reversal of Fortune," *Screen Digest* (September, 1994), 207.
58. Dodona, *Cinemagoing Scandinavia 2006* (Leicester: Dodona Research, 2006), 6; and Danish Film Institute, *Facts and Figures 2018* (Copenhagen: Danish Film Institute, 2018), 8. https://www.dfi.dk/files/docs/2018-04/Facts-and-figures-2018_0.pdf (accessed 4 January, 2019).
59. Gyory and Glas, *Statistics of the Film Industry in Europe*, 54.
60. Danish Film Institute, *Key Annual Numbers*, https://www.dfi.dk/en/english/numbers-and-statistics/annual-key-numbers (accessed 6 January, 2019).

61. Danmarks Statistik, https://www.statbank.dk/BIO9 (accessed 5 November, 2018).
62. London Economics/BIPE, *White Book of the European Exhibition Industry: Vol 1—Annex 2: Denmark* (1994) http://www.mediasalles.it/whiteboo/wb1_andk.htm (accessed 4 January, 2019).
63. Dodona, *Cinemagoing Scandinavia 2003* (Leicester: Dodona Research, 2003), 12 and Danmarks Statistik, https://www.statbank.dk/BIO9 (accessed 5 November, 2018).
64. "Danish Nordisk Emphasizes Wider Enterprises in Competitive Market," *Variety* 328, no. 13 (12 October, 1987), 394.
65. See Turner, *MGM Cinemas*, 1998.
66. Dodona, *Cinemagoing Scandinavia 2003*, 15.
67. See Chap. 9 under "Germany" for a detailed discussion of the company.
68. Dodona, *Cinemagoing Scandinavia 2009* (Leicester: Dodona Research, 2009), 5.
69. Dag Asbjørnsen and Ove Solum, "Public Service Cinema? On Strategies of Legitimacy in Policies for Norwegian Cinema," *International Journal for Cultural Policy* 5, no. 2 (1999): 269–91, 270.
70. Ibid., 280.
71. Ove Solum, "The Rise and Fall of Norwegian Municipal Cinemas," in *A Companion to Nordic Cinema*, eds. Mette Hjort and Ursula Lindqvist (Chichester: Wiley Blackwell, 2016).
72. See Petra Buddrus, "Nordic Giants Jostle for Position," *Screen International* 1222 (20 August, 1999), 24.
73. Dodona. *Cinemagoing Scandinavia & the Baltic, Update 2001* (Leicester: Dodona Research, 2001), 138.
74. Marlene Edmunds, "Exhibs Making Inroads," *Variety* 373, no. 12 (8 February, 1999), 67.
75. Dodona, *Cinemagoing Norway 2014* (Leicester: Dodona Research, 2014), 3.
76. Dodona, *Cinemagoing Scandinavia & the Baltic 2017* (Leicester: Dodona Research, 2017), 24.
77. ODEON Cinema Group, *World Class ODEON Opens in Storo, Oslo*, 23 March, 2018. https://www.odeoncinemasgroup.com/news/world-class-odeon-cinema-opens-storo-oslo/ (accessed 16 January, 2019).
78. Jorn Rossing Jenson, "Environmentally Sound," *Screen International* 734 (9 December, 1989), 38.
79. Ibid.
80. "Stockholm's Filmstaden Adds Four New Cinemas, Aims for a Million Admissions," *Screen International* 471 (10 November, 1984), 10.
81. Dodona, *Cinemagoing Sweden 2014* (Leicester: Dodona Research, 2014), 5.

82. See Keith. J.R. Keller, "Major Swedish Exhib Chains Undergo Multiplex Conversion," *Variety* 293, no. 9 (3 January, 1979), 24.
83. Marlene Edmunds,. "Hardtop Wars," *Variety* 366, no. 3 (17 February, 1997), 56.
84. Marlene Edmunds, "Nordic Building Spree Continues," *Variety* 371, no. 3 (15 June, 1998), 66.
85. Edmunds, "Hardtop Wars," 56.
86. Quoted in Marlene Edmunds, "AMC plans Swedish 24-plex," *Variety*, 14 October, 1997. https://variety.com/1997/film/news/amc-plans-swedish-24-plex-1116674243/ (accessed 3 November, 2018).
87. Dodona, *Cinemagoing Sweden 2014*, 158.
88. Jorn Rossing Jenson, "Swedish distrib Svensk Snubs New AMC Multiplex," *Variety* 382, no. 8 (11 April, 2001), 53.
89. Andreas Fuchs, "AMC is #1 After Nordic Acquisition," *Film Journal*, 14 April, 2017. http://www.filmjournal.com/columns/amc-1-after-nordic-acquisition (accessed 2 November, 2018).
90. ODEON Cinema Group has three subsidiaries: Odeon, UCI, and the Nordic Cinema Group. Odeon-UCI was purchased by AMC in 2016 (see Chap. 6) and subsequently split into two, with Odeon being the British arm and UCI the European Arm.

# CHAPTER 9

# The Multiplex in Germany and France

Germany and France make contrasting case studies of the multiplex's development in Europe, particularly when one considers the efforts of US exhibitors to enter those two markets. Both were some way behind countries like Belgium, Sweden, and Britain in their adoption of the multiplex, whilst both had dissimilar domestic film exhibition structures and ownership patterns, especially with regard to the amount of protectionism afforded to those domestic industries. Both France and Germany had different planning regulations and priorities when it came to approving the construction of large multiplex cinemas. Finally, the extent to which US-produced films had penetrated the relative markets is also significant. In 1990, on the eve of multiplex developments in Germany, 83.8 per cent of films shown in cinemas were US-produced.[1] In France, in 1992, just before that country's first multiplex opened, the comparable market penetration of US-produced films was lower at 58.1 per cent.[2]

## GERMANY

From the beginning of the 1950s, admissions to West German cinemas (including those in West Berlin) increased enormously, with a peak in 1956 of 817.5 million admissions.[3] The decline in cinemagoing thereafter was continuous until 1976 when 115 million people visited the cinema, as attendance fluctuated from this point on with a peak in 2001 of 178 million admissions, before declining to 122.3 million visits in 2017.[4] Given that the totals for admissions post-1990 include the former East Germany,

© The Author(s) 2019
S. Hanson, *Screening the World*,
https://doi.org/10.1007/978-3-030-18995-2_9

then for a country of approximately 82 million people this is a very low figure. Indeed, it constitutes 1.5 visits per year compared to 1956 when Germans attended on average 15 times.[5]

Whilst neighbouring Belgium, for instance, saw no US or British multiplex companies expanding into its market the same has not been the case in Germany; which represented a market in which US companies opted for a more aggressive policy since it was an area in which significant growth was foreseen. Despite a greater screen density than in Britain, US companies speculated that as both a rich country and one in which the population spent less on cinemagoing—$0.20 for every $1 spent in the USA—Germans could be persuaded to go the cinema more often.[6] As far back as 1986, in the context of the opening of several multiplexes in Britain, Cinema International Corporation (CIC) was proposing building similar complexes in what was then West Germany. It asked Paramount Filmproduction Germany's head Willi Benninger, as a consultant to CIC International Theatre Group in London, to scout locations in shopping complexes in cities such as Frankfurt though few, if any, such locations were formally identified.[7] *Variety* noted though that CIC's proposals ran counter to the prevailing trend in terms of attendance, which in 1985 had dropped 8 per cent from the previous year.[8] Indeed, like Britain, attendance at West German cinemas had been declining year on year, with 142 million attending just six years before, in 1979.[9] Nonetheless, it was a clear statement of intent. Much of the speculative multiplex planning was focussed on West Germany, prior to formal reunification in October 1990. The move into the "New Lander" (the former East Germany) was also a significant feature of the growth in multiplexes in Germany, particularly from the mid-1990s onwards. In 1991, according to Dodona, 13 million former East Germans visited the cinemas once a year on average, which had increased to 27.7 million by 1999.[10]

As in Britain, much of the initial investment was by US-owned companies. In 1989, United Cinemas International (UCI), which had been newly formed with the sale of American Multi-Cinema's (AMC) cinemas to CIC and United Artists (see Chap. 5), announced formal plans to expand into the West German market with proposed multiplexes in Bochum and Cologne, to be completed by late 1990 and a further 12 complexes across the country.[11] As was the case in Britain, UCI's Millard Ochs said that the sites would all be within a 20-minute driving distance of populations of 300,000 or more.[12] Ochs' colleague, Bernd Rohle, said that UCI had plans "to build a multiplex in every German city."[13] In the same year, Warner Bros. International Theaters (WBIT) announced it would be building a nine-screen multiplex in Gelsenkirchen, on the site of a former drive-in cinema, with proposals for six more to follow. Both UCI and WBIT, which had consolidated their position in the rapidly expanding British market, felt that the model could be applied to Germany. Indeed, according to UCI's Wolfgang Braun the "pattern of multiplexes in the UK will be followed."[14] Nonetheless, Wolf-Dietrich von Verschuer, chief executive of the Association of German Cinemas, cautioned against direct comparisons with Britain especially with regard to simple correlations between increases in audiences and the building of multiplexes.[15] Von Verschuer highlighted Germany's large number of exhibitors (approx. 80), many of them single-screen operations, whilst the significant drop in admissions in Britain over the previous ten years had been reflected in Germany.[16]

If December 1985 was the 'moment of the multiplex' in Britain, then in Germany it was October 1990, with the formal opening of UCI's 14-screen complex in Cologne's Hurth Park shopping centre. Just a few months later in March 1991, WBIT, in collaboration with domestic company Constantin Kinoholding-Beteiligungs-Gesellschaft GmbH (Constantin Film), opened its Gelsenkirchen multiplex.[17] The two companies formed a joint company—Constantin-Warner Kino GmbH—in 1991 with plans to invest DM1.5 billion for 15 multiplexes, with WBIT responsible for suburban developments and Constantin for those in urban centres.[18] Both the area around Cologne and the Ruhr region quickly emerged as the epicentre of the multiplex revolution, as Warner-Constantin opened the 13-screen Cinedom in Cologne's Mediapark and UCI opened an 18-screen complex in Bochum, both in late 1991. Within 14 months of the first multiplex, there were six in Germany with 81 screens, with another four planned for 1992 and a further three for 1993. Throughout the 1990s 128 multiplexes were opened with 1162 screens.[19]

As in the first ten years in Britain, the German multiplex market experienced what Williams and Hills described as an "incipient boom," with half of the country's 30 cinemas with nine screens or more having opened in 1996 alone.[20] It was boom that was based upon a reversal of declining attendance, which had risen albeit erratically from 1990, though it was also one which peaked, as we have seen, in the early 2000s. In 1991, admissions to traditional cinemas were 97 per cent, which had fallen to 59 per cent by the end of the decade, when the proportion of multiplex screens was 25 per cent of the total.[21] Unlike Britain and other European countries, however, Germany's cinema market has, according to Dodona, shrunk rather than expanded since 2001, with the 2017 total for admissions of 122.3 million less than that of 1993.[22]

As we have seen, the boom was initially led by US companies, operating as arms of their British operations, but as the market matured a series of domestic exhibitors emerged, some of which moved into the multiplex sector from traditional cinema operations, whilst others were the result of joint ventures with overseas operators. Several large international multiplex operators, from the USA (WBIT) and Australia (Hoyts[23] and Village Roadshow), pulled out of the German markets selling their operations to other exhibitors, as they did in other European countries largely because of financial difficulties experienced by parent companies in their home territories. A number of domestic operators emerged, notably CineStar, CinemaxX, and UCI Kinowelt. Unlike Britain, however, which was characterised by a high concentration of ownership by 2003 with six companies controlling 80 per cent of all screens,[24] in Germany in the same year the six largest circuits controlled only 27 per cent of screens.[25]

By the end of the first decade of the twenty-first century, the largest exhibitor in Germany was the Australian company Greater Union, which had begun as a joint venture with Heiner and Marlis Kieft, who inherited the family cinema business began in Lübeck after the Second World War. The company opened its first multiplex in Lübeck in 1993, followed by several others under their CineStar brand, with a particular focus on the newly emerging market in the former East Germany. Unlike many of their US competitors, the company focussed, though not exclusively, on smaller towns and cities and often built in the centres rather than on the periphery. In 2003, Greater Union, funded by its parent Amalgamated Holdings Ltd, took control of the whole company and by 2010 had 554 screens (see Chap. 10). Amalgamated Holdings, which has increasingly retrenched back to its Australasian markets in recent years, changed the name of its German arm to Event/CineStar, though by 2018 it remained Germany's largest exhibitor with 426 screens.[26]

The second largest operator to emerge was CinemaxX, which was formed by Hans-Joachim Flebbe in the late 1970s but which expanded greatly after forming a partnership with media company Scriba & Deyhle. Their first multiplex was a 10-screen complex opened in Hanover in March 1991, followed by a 16-screen multiplex, at the time the largest in Germany, in the Ruhr city of Essen. In 1998, the company was floated on the German stock exchange and in 1999 Belgian multiplex company

Kinepolis Group NV purchased a 30 per cent stake in the new CinemaxX AG, with a view to creating a pan-European operator building on the two companies' dominant positions within their domestic markets. Film producer and distributor Senator Films purchased a 25 per cent stake in the company in 2000 and in the same year CinemaxX took a 25 per cent stake in rival circuit UFA-Theater AG. This created the largest cinema circuit in Germany with 92 sites and 570 screens.[27] However, the new company's performance in this new pan-European venture was mixed and, like many German exhibitors, it dispensed with many sites in the 2000s, as audiences fell, and made substantial losses in the immediate period after taking over UFA-Theater AG; whereupon its main creditor, Deutsche Bank, had to write off substantial loans to the company.[28] In 2004, the company was effectively rescued by TeleMunchen in a deal that exchanged debt for equity. Kinepolis sold its stake in the company in 2009 and in 2012 it was taken over by British-based exhibitor Vue. Though it is still Germany's second largest exhibitor with 426 screens the newly renamed Vue/CinemaxX is substantially smaller than it was when it acquired UFA-Theater AG.

The third biggest and also the largest wholly foreign-owned exhibitor is UCI Kinowelt, the brand name for United Cinemas International Multiplex GmbH. As in Britain, when the company was formed in 1989 (see Chap. 5), UCI had been a key driver in the introduction of the multiplex into the German market, opening Germany's first purpose-built multiplex in 1990 in Cologne. In the 1990s, the company, having consolidated its British operations, looked to expand globally, with Germany indicated as a key area of growth. UCI's Alan McNair said they were looking for markets where they could be a "significant player" where "we'll be the first or one of the first," adding that the "trick is to convert those countries which are deficient in number of cinemas into cinemagoing populations."[29] By the end of the 1990s, UCI had opened 19 multiplexes with 175 screens and were one of the first companies to expand into the former East Germany, opening the first multiplex there in Leipzig in 1992. Despite their market position, UCI Kinowelt were constantly rumoured to have been unprofitable and in 2004 all of UCI's European operations were purchased by British-based private equity company Terra-Firma, as part of its Odeon/UCI group.[30] Regardless of such worries and occasional talk of divestment, the parent company has continued to see Germany as an important market and, whilst it has reduced the number of screens since 2010, it has also continued to open multiplexes, such as a 14-screen site in the Mercedes Platz complex in Berlin in 2018.

As in Britain, there were tensions between operators and local authorities over the desire on the part of operators to build the new multiplexes on the edge of towns and cities, on greenfield sites, and local authorities seeking to bolster their town and city centres. Volker Reich, chair of UFA-Theater AG, set out the argument clearly:

> The multiplex boom in Germany cannot be ignored. Even cities with previously poor attendance records have become highly coveted locations for multiplex cinemas. The major cinema operators, competing for these locations in Germany, are re-encountering each other in almost all the cities while competing for various construction sites. Discussions are presently raging in the municipalities regarding town planning: Should a new movie palace be allowed into the inner city, where it will be surrounded by age-old building, and where problems of traffic have to be resolved? Or should it be located outside the gates of the city "on a great belt" where there are adequate parking facilities, especially as problems of land ownership need not be clarifies and land is anyway much less expensive?[31]

Herbert Strate, head of the National Association of German Filmtheatre Owners (HDF), felt that the development of the multiplex was "a threat to us all" with the distinct prospect of "an American situation whereby 10 cinema circuits control 55% of the market."[32] For the established exhibitors the "apocalyptic vision" was "four or five hundred multiplexes placed at strategic locations in a highly motorised Germany."[33] Strate and others were also mindful of what was widely believed to be the unfair financial advantage that the US-owned companies had in making investment decisions. In short, exhibitors like UCI and Warner Bros. could draw on the resources of their parent companies to guarantee returns over periods of 20 years to developers, whilst local exhibitors' business plans were based solely on the returns from admissions predicted over a year or so.[34] Since 1991, the number of traditional screens has declined by around 215 at the same time as the number of multiplex screens has risen from 81 to 1413.[35] Notwithstanding the fears of traditional exhibitors then, Dodona suggested that "multiplexes have established themselves in the German market without dominating it, since most cinemas are still situated in town centres, meaning that the multiplexes cannot offer the free parking available in out-of-town sites."[36] As admissions in Germany have fluctuated since 2001, attendances at multiplexes and traditional cinemas have experienced a similar drop of around a third, whilst the proportion of screens accounted for by multiplexes, which looked like passing 50 per cent after 2010, declined thereafter to reach 44.5 per cent in 2017.[37]

In Germany, there was no specific legislation which applied to the construction of multiplexes, irrespective of location. What in Britain were known as "out-of-town" sites were the preferred option for the first wave of multiplex developments in Germany, not least because whilst any building had to conform to any land use restrictions currently in force,[38] they were less constrained by zoning codes and regulations, along with city ordinances. A freestanding multiplex was also cheaper, not least because it did not require architectural competitions. One of the corollaries of this lack of specific rules is that multiplexes have been built in the same catchment areas, with particularly regions seeing the greatest competition. In Dresden in 1997, for instance, UCI were granted planning permission for a 9-screen multiplex in Elbe park, three kilometres from the city centre, and which was completed before rival UFA-Theater AG had finished negotiating for a 15-screen complex in the city centre.[39] However, as the companies began expanding into the former East Germany in particular, there was a drive on the part of local authorities to encourage cinema building in town and city centres. This again echoed the experience of Britain with its shifts in planning policy after 1995 (see Chap. 7). In Germany though, as Ludemann argued, this was less of an issue for developers since German city centres were "well stocked with other ancillary facilities for eating, drinking and nightclubbing."[40] According to Dodona, where a multiplex opened in a city centre there almost always followed a significant increase in admissions, with Dortmund an example, as the opening of its multiplex in 1997 boosted the numbers visiting the cinema by over 200 per cent in a year.[41]

In the period between 2010 and 2017, admissions to Germany's cinemas have declined by 9.4 per cent as the number of cinema screens has risen by 3.8 per cent.[42] Though a number of significant domestic exhibitors emerged in the 1990s and 2000s, the three largest were subsequently bought out, increasingly by other foreign exhibitors, so that by 2018, the three largest exhibitors, which dominate the multiplex market in Germany—Event/CineStar, Vue/CinemaxX, and UCI Kinowelt—are foreign-owned.

## France

Unlike Germany, there are no US-owned (or indeed British-owned) multiplexes in France, though the multiplex as a form of cinema is dominant. In France, US companies found expansion difficult, not only because of its established antipathy towards Hollywood but also because of the strength of France's domestic exhibition industry, which like the USA had a high degree of vertical integration. In 1998, AMC's 20-screen megaplex proposed for the western suburb of Toulouse, but subsequently never built, resulted in a petition, organised by a local exhibitor, which warned local residents that "we have four months to stop this neo-colonial horror."[43] In 1999, AMC did open a 20-screen megaplex in Dunkirk, which had taken six years and would be its (and any other US exhibitors') only site. According to company head Bruno Frydman, the local press had billed it as "The Phantom Menace."[44] Frydman commented that "maybe I'm paranoid but I have a feeling we have been demonised."[45] Nevertheless, at the point at which the new shopping and leisure complex was being proposed for the run-down former port site of Dunkirk, none of the established French exhibitors was interested in operating the multiplex. Gaumont had an established multiplex in nearby Calais, which it was reluctant to compete with, whilst Union Générale Cinématographique (UGC) and Pathé claimed that the finances of the project were unattractive.[46] Though he had the support of the city officials, AMC's Frydman did have to counter opposition from local arthouse exhibitor Studio 43, which was overcome only after AMC agreed that the independent would show popular non-French films with subtitles whilst the AMC could show only dubbed versions.[47]

Cinema admissions in France had peaked in 1947 with 424 million visits,[48] whereupon they fluctuated with a further peak of 411.7 million visits in 1957, before they declined steadily until 1990 when they were 121.5 million.[49] In the period since then attendance has, like many territories, been generally upwards, albeit with fluctuations, and in 2018 there were 200.5 million admissions.[50] France is thus the number one cinemagoing country in terms of admissions in Europe, ahead of Britain, though its box-office gross is second to Britain's, due in part to lower ticket prices.[51] The decline in the number of screens over the period from the mid-1950s to 1993 was less marked than the drop in admissions, with 5732 in 1957 dropping to 4297 in 1992—the eve of the first multiplex—though the halving in the number of seats per screen suggests that many cinemas were subdivided and thus many closures took place.[52] It was not until August 1993, eight years after Britain, that the first purpose-built multiplex was opened by Pathé on the outskirts of Toulon. Thereafter, the expansion in screen numbers was almost entirely accounted for by multiplex developments but critically, these have been dominated by French-based exhibitors.

The multiplex's impact seemed dramatic since in the first three years, in which some 22 complexes were built, a study by the Centre National de la Cinématographie in 1997 revealed that 58 per cent of cinemagoers said that they had started attending more often since the multiplex began, whilst 78 per cent commented on the availability of parking, choice of films, and better technical facilities meant that they preferred visiting the multiplex rather than other forms of cinema.[53] As Dodona pointed out in 2013, 20 years after the first multiplex opened: "[t]raditional cinemas more or less held their own against the tide of multiplexes until around 2000 … Nevertheless, traditional cinemas still play an important role in city centres and small towns, and France still has 1,181 single screen cinemas."[54]

When the first multiplex was opened three companies dominated exhibition: UGC (Union Générale Cinématographique), Pathé, and Gaumont, each of which was also a vertically owned company with interests in production and distribution. Between 1993 and 2001, when Pathé and Gaumont merged their exhibition interests to form EuroPalaces, 1163 multiplex screens were built, comparable with Germany but well behind developments in Britain.[55] However, the key difference between France and both Germany and Britain was the almost total absence of US-owned multiplexes: indeed, AMC's Marine complex in Dunkirk was the only one. The only substantial foreign-owned operator was the Belgian Kinepolis Group, which had opened its first multiplex near Metz in 1995 and, by the end of 2001, operated seven multiplexes.

The complexion of the French exhibition sector, in terms of international penetration, had not differed by 2017, with Kinepolis still the only foreign exhibitor in the market. Remarkably, the exhibitors with the widest portfolio of sites across Europe—ODEON Cinema Group, Cineworld, and Vue[56]—have no presence in the French market. The context for this is a complex tension between an opposition to multiplexes on the grounds that they presented an existential threat to "traditional" cinema—both cinemas themselves and the projection of a distinct French national identity—and represented a form of Americanisation, which Kuisel identified as the "self-serving process of globalization" provoking what he characterised as a "neo-anti-Americanism."[57] In France, the development of the multiplex cinema has been framed then by a series of discourses around Americanisation, globalisation, and the importance of French national culture and identity.

At the heart of the debate was the health or otherwise of the town and city centre, in terms of both retailing and leisure, in the face of the out-of-town shopping and leisure complex. In July 1996, Jacques Chirac's government passed the Law Raffarin (hereafter Raffarin Act),[58] named after its proponent Jean-Pierre Raffarin, the minister for small business, which sought to tighten the regulations governing large-scale retail establishments. It was a development of 1973 legislation called the Law Royer that was designed to restrict the development of large-scale supermarkets and hypermarkets. The Raffarin Act, which adopted the slogan "A New Lease of Life for the City," created a two-tier system of regulation applied to cinema developments in which new builds of 1500 seats or more (later reduced to 999) and extensions to existing cinemas of 2000 seats or more would be examined by a departmental commission called the CDEC (Commissions départementales d'equipement commercial cinematographique).[59] A second national commission was created to examine permits awarded or refused and could overrule the local commission. This attempt to use legislation as a way of defending small retailers and independent cinemas in town and city centres and discouraging developments on the periphery was being echoed in Britain too, with the publication of revised planning legislation under John Major's Conservative government (see Chap. 7).

In an influential article in *Cahiers du Cinema* in 1997 on the perceived impact of the multiplex on independent cinemas, Carlos Pardo characterised contemporary cinema as a "banal product for consumption" with the multiplex deeply implicated in this by exhibiting predominantly mainstream films whose success is guaranteed.[60] According to Dodona, the regulatory structure in France slowed multiplex building but also inadvertently meant that overbuilding was less common and the sector has developed in a more sustainable and profitable fashion.[61] This regulatory structure and in particular the influence of localities have meant that multiplex development in France has often favoured town and city centres, with an emphasis on their role in urban regeneration schemes. As Hayes observed, this has meant that municipal authorities have played a role as "key entrepreneurs" as opposed to the more privatised initiatives that have characterised multiplex developments in Britain and Germany.[62]

When it comes to ownership, France has a less concentrated structure of ownership, with the three largest operators selling less than half of all tickets; markedly different from Britain where the figure is over 70 per cent (see Chap. 7). The largest operator currently is Les Cinémas Gaumont Pathé, which was formed with the merger of the exhibition arms of Gaumont and Pathé in 2001, whereupon the company was called

EuroPalaces, being 66 per cent owned by Pathé and 34 per cent by Gaumont. In 2011, the company changed its name to the current configuration, and in March 2017 Gaumont sold its stake to Pathé, in order to concentrate on TV and film production. Gaumont is vertically integrated with interests in production and distribution, alongside cinema operations in the Netherlands, Switzerland, Belgium, and Tunisia. The company's origins lay in the opening by Leon Gaumont of the 5000-seat Gaumont Palace in Paris in 1911—at that time the world's largest cinema. Gaumont was vertically integrated from the earliest days, with its exhibition arm an integral facet of the company, though in 1970 Gaumont had signed a programming pact with rival Pathé—known as the Groupement d'Intérêt Économique (GIE)—which saw them control 150 cinemas and 150,000 seats, largely, they claimed, to make distribution of their films more efficient. The pact with Pathé was finally outlawed by France's Socialist government in 1983 as a result of the film reforms instigated by the Minister of Culture, Jack Lang. In November 1981, Lang's ministry had published a report, compiled by a commission of film industry professionals chaired by Jean-Denis Bredin, which had recommended the breaking up of the Gaumont-Pathé pact, since it controlled more than 30 per cent of first-run cinemas, and sought more competition between the three major vertically integrated companies: at that point Gaumont-Pathé, Parafrance,[63] and UGC.[64] Moreover, Lang's reforms also sought to separate out the booking activities of these companies from their distribution and exhibition arms, mirroring the 1948 *Paramount Decrees* in the USA.[65]

Like its US counterparts, Gaumont's development before and through the multiplex era was tied up with a charismatic head Nicolas Seydoux, who bought a controlling stake in the company in 1975. At that time, like the major British exhibitors, Gaumont had subdivided many of its cinemas in an effort to make them more economically viable. However, after the ending of the pact with Pathé in 1983, and in the face of a steady decline in attendance, Seydoux instigated a major revamp of the circuit including the acquisition of 100 new screens and a costly renovation of several prime sites, which was dubbed GaumontRama. The renovations included larger screens, and better sound and projection. Significantly, Seydoux did not recommend the building of new multiscreen cinemas, on suburban and greenfield sites, insisting that the Gaumont audience "prefers a more cosmopolitan, in-town environment: thus, city center sites are being earmarked for future development."[66]

The investment in GaumontRama into the 1990s was substantial and it resulted in a renewal of the agreement, of sorts, with Pathé, owned by Nicolas' brother Jérôme. This saw the two companies exchange 66 cinemas in 1992, with Gaumont gaining a larger presence in Paris at the expense of regional cinemas in cities like Nice, Toulon, and Grenoble. Both companies insisted that the arrangement would allow them to invest in cinema upgrading which was, according to Pathé, "an essential condition in facing up to the fall in cinema attendances."[67] Pathé in turn made a smaller swap with rival UGC. Moreover, by the mid-1990s and in the wake of Pathé's opening of France's first multiplex, Gaumont instigated a multiplex building programme which saw its first multiplexes—12-screen complexes in the centre of Nantes and near the Eurotunnel terminal in Calais—opened in 1995. Just prior to the merger with Pathé, Gaumont operated 16 multiplexes with 204 screens.[68]

Pathé's origins go back as far as the cinema itself when in 1896 with brothers Emile, Théophile, and Jacques, Charles formed the Pathé-Frères company. They opened their first cinema—the Omnia-Pathé in Paris in 1906—and by 1910 Pathé owned some 200 cinemas in both France and Belgium. Like Gaumont, Pathé have been a vertically integrated company with interests in film stock and equipment manufacture, production, and distribution; whilst their exhibition arm grew significantly over the period. The ending of the Gaumont-Pathé pact in 1983 meant that Pathé became the third largest circuit in France but also coincided with a downturn in cinemagoing in France. Pathé's then head Pierre Vercel argued that this was due in part to an economic crisis and a dearth of American films (exacerbated by the 1983 reforms which meant that Gaumont and Pathé could not show the same film at the same time in the same area).[69] Pathé looked to increase the profitability of its cinemas and lure audiences back in the late 1980s via the introduction of a computerised advance booking and a subscription system which it called "cinema à la carte."

At the end of the 1980s, Pathé was the subject of a bruising and protracted attempted takeover by Italian financier Giancarlo Parretti, who purchased 46.5 per cent of the company in 1989.[70] The French Government effectively blocked his attempt to acquire the remaining stock from the Anglo-Belgium Rivaud Group, with the company passing into the ownership of the Chargeurs Conglomerate, controlled by the Schlumberger family, in 1990.[71] The exhibition circuit, along with the company's other media interests including a significant stake in British satellite broadcaster BSkyB, was subsequently spun off as a new company called Pathé SA in 1996, headed by Jérôme Seydoux. Like his brother Nicholas at Gaumont, Jérôme set about modernising the company's cinemas as soon as Chargeurs took over, instigating a programme it called Pathé Rama, including new seating, sound, lighting, and decoration, in addition to what it called a new "Pathé Identity" in the form of a new logo. Pathé were also able to capitalise on the first significant rise in admissions in 1990 and which grew across the subsequent decade.

The opening of the first multiplex by Pathé—the Pathé Grand Ciel—in La Garde, a suburb of Toulon, in June 1993 marked a turning point in exhibition in France. The 12-screen complex cost FFR87 million and in the first year of operation was expected to attract 800,000 admissions, which would have required those for the whole Toulon-Hyeres region to increase by 50 per cent so as not to have a deleterious impact on Pathé's existing 6-screen cinema in the city centre.[72] In reality, admissions in the region increased by 40 per cent as opposed to 15 per cent for the country as a whole.[73] Pathé set about an ambitious building programme with a further 12-screen multiplex—Pathé Belle-Epine—opened in Orly on the outskirts of Paris in October 1993 and a further 11 sites opened before the end of 2001, including a 16-screen complex—Pathé Plan de Campagne— in Pennes Mirabeau, Marseille, opened in July 1997. Harold Alsberge, Pathé's Vice President-Operations, referred to their two new multiplexes as "temples of cinema and leisure" that would reverse the decline in box-office, because spectators have "rediscovered the meaning of the word 'pleasure' and sought an 'escape to the world of fun'."[74] At the point in which Pathé merged with Gaumont to form EuroPalaces in June 2001, the circuit operated 13 multiplexes with 156 screens; moreover, it was the largest exhibitor in the Netherlands with 32 per cent of admissions from 18 cinemas and 105 screens (see Chap. 8).[75]

The second largest exhibitor in France after Les Cinémas Gaumont Pathé is Circuit Georges Raymond (CGR), which began as a regional company based in La Rochelle in the south-west of France and expanded into a national circuit due to its early adoption of multiplexes. Georges Raymond, who formed a media company in 1966, and who died in 2001, purchased his first cinema—the L'Olympia in La Rochelle—in 1974. In December 1995, CGR opened its first multiplex—a 12-screen complex— in La Rochelle under its brand name Méga CGR. This was followed by 11-screen complexes in Angouleme in May 1996 and Pau in October 1996. It generally opted for larger towns and smaller cities, initially in its heartland of the south-east but expanded across the country subsequently.

Georges Raymond set out the philosophy for multiplex expansion thus: "[F]or medium-sized towns, the multiplex has no 'raison-d'être'... it could completely empty out the town centre. The traditional movie complex, therefore, has its future there, on the condition it is adapted to the multiplex standard (large screen, high fidelity sound)."[76] In the four years after the opening of its first multiplex, CGR opened a further three or four per year until by 2001 it operated 20 multiplexes, more than established rival UGC.[77] The company continued its aggressive expansion throughout the period up until 2017, whereupon in November it acquired smaller rival Cap'Cinema and its 22 cinemas, including 12 multiplexes and two projects in Nantere and Grasse.[78] Significantly, CGR had increased its admissions over the period 2014–17, before it acquired Cap'Cinema, unlike its largest rivals Les Cinémas Gaumont Pathé and UGC.[79]

In 2008, CGR was the first exhibitor to convert all of its multiplexes to digital projection, having signed an agreement with aggregator Arts Alliance Media (AAM) in 2007 to equip 400 auditoria with Digital Cinema Initiative (DCI)-compliant 2K digital projectors. AAM had a pact with the five Hollywood studios which helped to fund CGR's transition via the Virtual Print Fee (VPF) model.[80] CGR saw digital technology as inevitable and its adoption partly concerned with capitalising on 3D cinema, and stimulating French distributors to change. Moreover, according to AAM's Gwendal Auffret, digital cinema would help CGR to optimise its auditoria, adding that "in the digital realm, they'll be more efficient by being able to move the features around."[81] The move to digital, which CGR did very quickly, was followed in 2016 by the development of Immersive Cinema Experience (ICE), which was based upon Philips' LightVibes technology, adding ambient video and surround lighting effects to viewers' peripheral field of vision. CGR combined it with Christie laser projection, Dolby Atmos sound, and reserved stadium seating, first equipping its Blagnac multiplex, followed by three more by 2017.

The third major French exhibitor, and one that has had a significant presence beyond France, is UGC, formed in 1971 when a group of independent exhibitors, led by Jean-Charles Edeline, the head of the Fédération Nationale des Cinémas Français (FNCF), purchased 22 cinemas off the French government for FFR60 million. The cinemas, including 6 in Paris and 16 in key cities, had been seized by the Germans during the Second World War and were then confiscated by the French government in 1945, which ran them through a company called Générale Cinématographique. The condition of the sale was that the circuit be maintained, with some suggestions that the sale was in part designed to keep the cinemas out of the hands of both domestic and foreign rivals, the USA in particular.[82] UGC expanded slowly under Edeline so that in 1974 when Guy Verrecchia took over as chairman, the company operated 45 cinemas and had approximately 30 per cent of the market behind then rivals Gaumont Pathé and Parafrance.

Through a process of acquisition and refurbishment, UGC became France's largest exhibitor by 1994, with 230 screens and 36 per cent of those in the Paris. With its purchase in 1992 of a multiscreen complex called the Forum Horizon from the Lemoine family, in what was the Forum des Halles shopping mall, UGC's concentration of ownership in Paris precipitated a nine-month investigation by French Culture Minister Jacques Toubon. The result was that UGC was forced to divest itself of a number of cinemas in the Montparnasse district and sell a site on the Champs Elysees. One of the outcomes of the investigation was that UGC would have to concentrate on building, rather than acquisition, in order to expand and thus embraced the concept of the multiplex in November 1994, with a 10-screen complex (later increased to 14 screens) in Lille. The cinema was distinguished from contemporary developments amongst competitors by its location in the town centre rather than on a greenfield site. Moreover, the company eschewed the generic term 'multiplex' in favour of its "signature cinema concept" Ciné Cité, with cyber cafes and large concession stand, and a focus on town and city centres, which Verrecchia identified as important when attracting teenagers and young adults.[83] The considerable cost of the planned country-wide Ciné Cité multiplex expansion was funded by a FFR700 million joint venture with the investment bank Lazard.

In June 1995, UGC opened what was then France's biggest multiplex, a 15-screen complex in the Les Halles district of Paris, which was in part the Forum Horizon along with an extension to the building on what had been the neighbouring Espace Cousteau, a former aquatic park. In 1997, the multiplex became a megaplex with the addition of four more screens. UGC claimed that the cinema also had the highest rate of annual admissions in France, with 2.4 million in 2000, whilst it was also the most visited arthouse cinema in France with 60,000 visitors.[84] Within six years of opening the Ciné Cité Lille, UGC had added a further ten multiplexes including an 18-screen megaplex in Bercy, Paris, opened in December 1998 and the 22-screen, 5400-seat Ciné Cité Strasbourg, opened in November 2000.[85]

Significantly, amongst the major French exhibitors in the 1990s, UGC looked to expand abroad. UGC had entered the Belgian marketplace in the late 1970s, becoming the second largest exhibitor after Kinepolis from 1988 (see Chap. 8). In Spain, UGC Iberia opened its first multiplex—the 16-screen Ciné Cité Madrid—in December 1997, and by 2011, when it sold its interests to Terra Firma's Odeon/UCI group, it operated five multiplexes. Similarly, UGC also expanded into Italy before selling its four multiplexes in the same deal. Most significantly, UGC expanded into the British multiplex market in 1999 when it purchased Virgin Cinemas (see Chap. 6). This was a substantial acquisition with UGC controlling 33 British and two Irish multiplexes, totalling 300 screens, and by 2001, UGC was the second largest operator in Europe, with 82 sites and 693 screens in five countries.[86]

Having drawn attention from the French Ministry of Culture in 1992, over concentration of ownership, UGC once again did so in 2000 when Minister Catherine Tasca recommended that UGC go before a board of fair trading as a result of the introduction of the *Carte UGC Illimitée*. On 29 March 2000, UGC began offering a pass that, for FFR98 a month, would give the subscriber unlimited access to UGC cinemas around the country. The idea was based upon an innovation of the recently acquired Virgin Cinemas and its *Megapass* (see Chap. 6). In the first two weeks, some 20,000 were sold, much to the ire of smaller exhibitors and various sections of the film industry,[87] with UGC having to suspend sales in May whilst the board investigated. UGC's Guy Verrecchia argued that far from trying to put smaller exhibitors out of business, the card was primarily designed to make the price perception between the cinema and television "painless."[88] The argument was that people tended to over-estimate the price of a cinema ticket and under-estimate the cost of television viewing. According to Bonnell, the unit price per visit was replaced by a cost similar to that of subscription to a pay-TV channel.[89]

The outcome of the investigation was that legislation was introduced so that UGC were cleared to offer the card again from July though they had to pay a fixed fee to distributors each time it was used. Within a few months, in August, Pathé launched its own card—*Cine a volante*—and in September 2000, Gaumont and arthouse exhibitor MK2 launched their version—*Gaumont-Mk2 le Pass*. In January 2001, legislation was introduced which meant that 'dominant exhibitors'—with 8 per cent of the Paris market or 0.5 per cent of the national market—had to open their schemes up to other exhibitors in the same zone.[90]

Though France's first multiplex did not open until eight years after Britain's first, by the end of the 1990s France was second only to Britain in terms of numbers of multiplex screens, and significantly ahead of its nearest rival Germany. Moreover, by 2017, whilst multiplexes accounted for 10.7 per cent of cinemas and 42.4 per cent of screens, they accounted for 60.1 per cent of admissions.[91] Significantly, the exhibition market in France is dominated by domestic companies, with only one foreign circuit, Belgium's Kinepolis, which has 128 screens in its multiplexes which is just 5 per cent of the total.[92] In 1994, several US- and British-owned companies were professing an ambition to enter the French market including AMC and UCI, though both had found it difficult to find local partners and instead were looking to strike agreements with financial institutions and investors.[93] As AMC's Bruno Frydman observed:

> When I was given the task of developing the AMC presence in Europe, I thought it would be simple: AMC had already established a European bridgehead in the UK in 1992 with a highly successful business, while the experience of some of our counterparts in Belgium and other European countries led us to expect a glowing future. We had announced plans to build a vast number of screens. Unfortunately, that's not the way it turned out.[94]

## NOTES

1. Martin Blaney, "Germans Fall Before US Box-Office Advance," *Screen International* 827 (4 October, 1991), 8.
2. Dodona, *Cinemagoing Western Europe 2004* (Leicester: Dodona Research, 2004).
3. SPIO, Spitzenorganisation der Filmwirtschaft e.V. *Kino Filmbesuch 1946– 2017, Filmbesuche in Deutschland 1946 bis 2017.* Wiesbaden. https:// www.spio-fsk.de/?seitid=381&tid=3 (accessed 26 August, 2018).
4. Ibid.
5. Ralf Dewenter and Michael Westermann, "Cinema Demand in Germany," *Journal of Cultural Economics* 29 (2005): 213–31, 214.
6. Ralf Ludemann, "To Protect and Nurture," *Screen International* 863 (26 June, 1992), 22.
7. "CIC Plans to Build German Multiplexes," *Variety* 324, no. 4 (20 August, 1986), 27.
8. Ibid.

9. "German-Speaking Market At a Glance—West Germany," *Variety* 314, no. 6 (7 March, 1984), 336.

10. Dodona, *Cinemagoing Western Europe 2000* (Leicester: Dodona Research, 2000), 89.

11. Mike Downey, "CIC, UCI Plan 12 German Plexes" *Screen International* 715 (29 July, 1989), 715.

12. Ibid.

13. Quoted in Patrick Frater, "Cinema Expo: Germany," *Screen International* 1013 (23 June, 1995), 25.

14. Mike Downey, "Multiplex Mania Hits W Germany," *Screen International* 735 (16 December, 1989), 1.

15. Ludemann, "To Protect and Nurture," 22.

16. Ibid.

17. In 1996, WBIT formed a joint venture with Australian exhibitor Village Roadshow International called Warner Village Cinemas Ltd, to own and manage its cinema operations in both Britain and Germany. WBIT's sites were rebranded as Warner Village Cinemas (see Chap. 5). In 1998, Village Roadshow bought out WBIT in Germany. See Mike Goodridge and Martin Blaney, "Warner, Village split in Germany," *Screen International* 1179 (9 October, 1998), 2.

18. Michael Blaney, "Constantin, Warner Plan Multis," *Screen International* 790 (18 January, 1991), 6.

19. FFA Filmförderungsanstalt, *Multiplexe (MP)—Besuch und Umsatz 1997 bis 2000 nach Generationen* (2000). https://www.ffa.de/download.php?f=3e1ee009355492b4f82f5d6637ffd69f&target=0 (accessed 27 August, 2018).

20. Michael Williams and Miriam Hils, "Local Euro Pix: Big Haul at Mall," *Variety* 366, no. 6 (10 March, 1997), 1 and 99.

21. Dodona, *Cinemagoing Western Europe 2000*, 102 and 95.

22. Ibid., 1.

23. Hoyts entered the German market via a 50 per cent stake in exhibitor Thiele Entertainment, which operated 18 sites with 162 screens in 1999, prior to selling its stake to the Kinowelt Arthaus Kino circuit in 2000.

24. Dodona, *Cinemagoing 12* (Leicester: Dodona Research, 2004), 30.

25. Dodona, *Cinemagoing Western Europe 2004*, 112.

26. Dodona, *Cinema Industry Research: Germany, Austria, Switzerland 2018* (Leicester: Dodona Research, 2018), 4.

27. Ed Meza, "Link-Ups Alleviate German Plex Crunch," *Variety* 379, no. 5 (19 June, 2000), 53.

28. Dodona, *Cinemagoing Western Europe 2004*, 119.

29. Quoted in Celia Duncan, "10 Years of UCI: If You Build It, They Will Come," *Screen International* 1026 (22 September, 1995), 29.

30. Dodona, *Cinemagoing Western Europe 2005* (Leicester: Dodona Research, 2005), 70.
31. Volker Reich, "A Multiplex in Every Large City," *The Film Journal* 98, no. 6 (1 July, 1995), 28.
32. Quoted in Jack Kindred, "Exhibition Cologne: Ruhring to Go," *Screen International* 823 (6 September, 1991), 26.
33. Ibid., 26.
34. Ralf Ludemann, "German Exhibition: The Space Race," *Screen International* 1095 (14 February, 1997), 32.
35. Dodona, *Cinema Industry Research: Germany, Austria, Switzerland 2015* (Leicester: Dodona Research, March 2015), 3. and Dodona, *Cinema Industry Research: Germany, Austria, Switzerland 2018*, 5.
36. Dodona, *Cinema Industry Research: Germany, Austria, Switzerland 2018*, 3.
37. Ibid., 3.
38. See Lothat Blatt, "Comparison of Regulations Governing the Opening of Cinema Complexes in Germany, France and Great Britain," MEDIA Salles round table *The Impact of Multiplexes on the Cinema Market and on their Environment*, Amsterdam—Cinema Expo International, 15 June 1998. http://www.mediasalles.it/expo98bl.htm#ing (accessed 25 October, 2018).
39. Volker Reich, "Booming Investments Vs. Laisser-Faire In Germany," *Film Journal International* 100, no. 6 (1 July, 1997), 34 and 36.
40. Ludemann, "German Exhibition: The Space Race," 30.
41. Dodona, *Cinemagoing Western Europe 2000* (Leicester: Dodona Research, 2000), 87.
42. Dodona, *Cinema Industry Research: Germany, Austria, Switzerland 2018*, 3.
43. Michael Williams, "Gauls to AMC: Yankee Go Home!" *Variety* 370, no. 10 (20 April, 1998), 14.
44. Francoise Maux Saint Marc, "Multiplex Moderation," *Screen International* 1241 (21 January, 2000), 16.
45. Ibid., 16.
46. John Tagliabue, "Now Playing Europe: Invasion of the Multiplex; With Subplots on Pride and Environment," *The New York Times*, 27 January, 2000.
47. Ibid.
48. Susan Haywood, *French National Cinema*, Second Edition (London: Routledge, 2005), 25.
49. Michel Gyory and Gabriele Glas, *Statistics of the Film Industry in Europe* (Brussels: The European Centre for Research and Information on Film and Television (CERICA), 1992), 97.

50. CNC, *Fréquentation des salles de cinéma : 200,5 millions d'entrées en 2018.* 31 December 2018. https://www.cnc.fr/professionnels/actualites/frequentation-des-salles-de-cinema%2D%2D2005-millions-dentrees-en-2018_903437 (accessed 13 January, 2019).

51. Melanie Goodfellow, "France Admissions Fell 4.3% in 2018 But It Remains Europe's Top Cinema-Going Territory." *Screen Daily* (3 January, 2019). https://www.screendaily.com/news/france-admissions-fell-43-in-2018-but-it-remains-europes-top-cinema-going-territory/5135539.article (accessed 14 January, 2019).

52. Dodona, *Cinemagoing Western Europe 2000*, 69.

53. See Michael Williams, "Plexes pack 'em in, Franco study finds," *Variety* 369, no. 2 (17 November, 1997), 22.

54. Dodona, *Cinemagoing Northern Europe 2013* (Leicester: Dodona Research, 2013), 18.

55. Dodona, *Cinemagoing Europe 2002* (Leicester: Dodona Research, 2002), 126.

56. As of 2018: Odeon Cinema Group (Odeon, UCI, and the Nordic Cinema Group) operated cinemas in Scandinavia, Britain, Ireland, Spain, Portugal, Italy, and Germany; Cineworld operates in Britain, Ireland, Poland, Hungary, Romania, Czech Republic, Bulgaria, and Slovakia; and Vue operates in Britain, Germany, Poland, Italy, and the Netherlands.

57. Richard Kuisel, "The Gallic Rooster Crows Again: The Paradox of French Anti-Americanism," *French Politics, Culture & Society* 19, no. 3 (2001): 1–16, 9.

58. Law No. 96-603 of July 6, 1996, Journal Officiel de la Republique Francaise J.O., 6 July, 1996.

59. See Graeme Hayes, "Regulating Multiplexes: The French State between Corporatism and Globalization," *French Politics, Culture & Society* 23, no. 3 (Winter, 2005): 14–33 and René-Paul Desse, "Les centres commerciaux français, futurs pôles de loisirs?" *Flux* 4, no. 50 (Octobre–Décembre, 2002): 6–19.

60. Carlos Pardo. "Multiplexes, opération danger (1)," *Cahiers du Cinéma* 514 (June, 1997): 60–67, 69. Cited by Hall. Katie. "France's Multiplex Gauntlet," *Film Journal International* 100, no. 8 (1 September, 1997), 34.

61. Dodona, *Cinemagoing Western Europe 2000*, 60.

62. Hayes, "Regulating Multiplexes," 19.

63. Parafrance began in 1963 and sold out its exhibition interests in 1986 amidst financial difficulties, when it was the third largest exhibitor with over 300 screens.

64. See Lenny Borger, "French Reformers Aim to Split Gaumont, Decentralize Exhibition; Push Local Role for U.S. Majors," *Variety* 305, no. 2 (11 November, 1981), 3.

65. See Ted Clark, "France To Ape U.S. Divorce, 1950 Of Distribs And Theatre Control," *Variety* 306, no. 10 (7 April, 1982), 33.
66. "Far From Resting on Its Laurels, Gaumont, at 90, Launches Prod'n, Distrib and Theater Revamp Plans," *Variety* 326, no. 4 (18 February, 1987), 4.
67. Quoted in: Michael Williams, "Indie Exhibs Fear Pathe Swap Fallout," *Variety* 246, no. 1 (20 January, 1992), 45.
68. Dodona, *Cinemagoing Europe 2002* (Leicester: Dodona Research, 2002), 107–9.
69. "Focus on France: Pathé—Keeping Out of Limelight," *Screen International* 445 (12 May, 1984), 73, 86.
70. See Chap. 5 for details of Paretti's purchase of the Cannon group in Britain and the establishment of Pathé Communications Corporation.
71. For a detailed overview of the company's operations see: Nick Bell, "Charging Forward," *Screen International* 884 (20 November, 1992), 11–14.
72. Patrick Frater, "Pathe Chief Demands Lower Film Rental: Seydoux unveils Pathe's Multiplex Expansion Plan," *Screen International* 912 (18 June, 1993), 7.
73. Patrick Frater, "Cinexpo European Focus: France," *Screen International* 963 (24 June, 1994), 50.
74. Harold Alsberge, "Pathe Emphasizes Customer Satisfaction," *The Film Journal* 97, no. 6 (1 July, 1994), 30.
75. Dodona, *Cinemagoing Western Europe 2000* (Leicester: Dodona Research, 2001), 73–4.
76. Quoted by Kate Combault, "Foreign Correspondence: Cinéma Francais," *Film Journal International* 103, no. 6 (1 June, 2000), 80.
77. Dodona, *Cinemagoing Europe 2002*, 146
78. *CGR Cinémas annonce l'acquisition du Groupe Cap'Cinéma et devient ainsi le 1er exploitant de cinémas en France*, CGR Press Release, 16 November, 2017. http://www.nextstage-am.com/wp-content/uploads/2017/11/Microsoft-Word-Communiqu%C3%A9-de-presse-CGR-Cin%C3%A9mas-Cap-Cin%C3%A9ma-Vdef_.docx.pdf (accessed 15 October, 2018).
79. Dodona, *Cinema Industry Research: France & Benelux 2018*, 11.
80. The Virtual Print Fee (VPF) model is a means of financing the conversion to digital cinema. A third party pays up front for the equipment, and then recoups the cost of the equipment over time, through payments from distributors (who pay the majority of the cost) and exhibitors.
81. Doris Toumarkine, "In the European Avant-Garde: CGR Cinemas Pioneers Digital Projection in France," *Film Journal International* 111, no. 7 (18 June, 2008), 76.

82. "Indie Exhib Group Buys French Govt. Circuit (22) For $11,850,000," *Variety* 261, no. 2 (25 November, 1970), 2.

83. Patrick Frater, "UGC Goes to Town on Multiplex Plans," *Screen International* 893 (5 February, 1993), 6.

84. Kate Combault, "Cinema Expo Honors UGC Co-founder." *Film Journal International* 103, no. 7 (1 July, 2000), 120.

85. This is the second largest multiplex in France after Kinepolis' 23-screen Lomme site.

86. "Uneven Pace of European Cinema," *Screen Digest* (September, 2001), 277.

87. Alison James, "UGC pic pass raises Gallic ire," *Variety* 379, no. 5 (19 June, 2000), 54.

88. Combault, "Cinema Expo Honors UGC Co-founder," 120.

89. René Bonnell, *La vingt-cinquième image: une économie de l'audiovisuel*, 3ᶜ édition: Gallimard, (2001), 154.

90. Dodona, *Cinemagoing Western Europe 2001*, 136.

91. CNC (Centre National du Cinéma et de l'Image Animée), *Results 2017: Films, Television Programs, Production, Distribution, Exhibition, Exports, Video, New Media* (May, 2017), 69. https://www.cnc.fr/web/en/publications/results-2017_557488 (accessed 23 September, 2018).

92. Dodona, *Cinema Industry Research: France & Benelux 2018*, 11.

93. Kim Williamson, "Yankee Go Home?" *BoxOffice*, 130, no. 7 (1 July, 1994), 30.

94. Bruno Frydman, "Exporting the Multiplex Model to Europe: The Experience of AMC" at the MEDIA Salles round table. *The Impact of Multiplexes on the Cinema Market and on their Environment*, Amsterdam: Cinema Expo International, 15 June, 1998. http://www.mediasalles.it/expo98fr.htm (accessed 10 September, 2011).

# PART IV

# The Multiplex Cinema in Asia-Pacific

# CHAPTER 10

# The Multiplex in Australia

Australia was one of the first countries to develop the multiplex and the maturity of the Australian market is perhaps second only to that of the USA. Indeed, Given states that the "connections between cinemas in Australia and the United States are long and deep. Australia has often followed American trends, occasionally led and has always linked closely to them."[1] Australia makes a fascinating case study of the development and diffusion of the multiplex cinema, not least because companies like Hoyts, Village Roadshow, and Greater Union have utilised their established market dominance in Australia to export the multiplex and influence developments across the world, via aggressive and extensive expansion into Europe, the USA, and Asia-Pacific.

As in the USA and Britain, the peak of cinemagoing in Australia was in the immediate post-war period, when there were 151 million visits in 1945, which then declined year on year until the nadir in 1977 when there were just 24.1 million visits.[2] The decline between 1976 and 1977 was an unprecedented 36 per cent, due in part to economic recession and the widespread purchase of colour televisions in major population centres like Melbourne and Sydney.[3] Thereafter, the number of admissions rose, albeit erratically, until 2001, when it reached 92.5 million before fluctuating between 82 and 92.5 million.[4] In 2002, Australia had the highest rate of cinema attendance—4.7 admissions per person—than any other country, though by 2017 this had dropped to 3.5.[5] In terms of cinema screens, the trend was for a decline in the number of screens from the 1950s onwards.

© The Author(s) 2019
S. Hanson, *Screening the World*,
https://doi.org/10.1007/978-3-030-18995-2_10

In 1951, there were approximately 2072 screens in Australia, which dropped by over a quarter to approximately 1481 in 1961 before halving to some 733 by 1971.[6] The number of screens reached its nadir of 645 in 1987 before beginning an inexorable and rapid rise until around 2000, whereupon it slowed down somewhat, reaching 2210 in 2017.[7]

## THE PRE-CURSORS OF THE MULTIPLEX

The "multiplex era"[8] is generally acknowledged to have begun with the opening in December 1986 of an eight-screen, 2750-seat site in the redeveloped Chadstone Shopping Centre in the southeast of Melbourne.[9] It was built by Hoyts, in a co-venture with Cinema International Corporation (CIC), the exhibition arm of MCA Inc. and Paramount Communications, and was firmly inspired by the experience of the USA. As Hoyts' General Manager Paul Johnson observed, the company "saw what was happening in the US and it was seen to be good for everybody. We thought it would transfer to Australia, and as screen numbers had been dropping, there was a need emerging."[10] Johnson's implicit optimism about the latent demand seemed justified, since by the end of the first year the Hoyts Chadstone had admitted over one million people.[11] In its first week of operation, some 20,000 people visited the multiplex.[12] As in the USA and Britain, the location—a shopping centre, in the suburbs, with mass free parking and a surrounding population of more than 200,000—had been carefully selected, with Hoyts and CIC announcing that they would be building a total of ten suburban multiplexes within four years.[13] The criteria for site selection were a minimum population of 100,000, the absence of existing competition, and an area in which all theatrical products was available.[14]

Chadstone was not the first large, purpose-built, multiscreen cinema in Australia however, since in December 1976, the seven-screen Hoyts Entertainment Centre had opened in the centre of Sydney.[15] Hoyts billed the Entertainment Centre as the "biggest cinema complex in the world" and claimed that in the first 18 months more than three million people had been admitted.[16] The 4300-seat cinema, or "Septplex," was in a building that also included a discotheque, a bar, and a delicatessen, with Hoyts arguing that "no longer is the mere presentation of a film enough to cause people to go to the cinema. We've got to give them significantly greater attractions than the old houses offered."[17] Its significance as the forerunner of the multiplex cinema was in one of the rationales for the multi-screening philosophy, which the company identified as the facility to drop the kids off to watch a G-rated film whilst the parents went in to watch an R-rated one, picking up the kids afterwards.[18]

As in the USA, exhibitors in Australia had presaged developments like the Hoyts Entertainment Centre, by subdividing existing single-screen cinemas but more significantly by building twin- and triple-screen cinemas. Village Cinemas began opening a series of twin cinemas in the 1970s. In New Farm, Brisbane, under its brand Village Twin Cinemas, it commissioned cinema architects Ron G. Monsborough and Associates to refit the existing Astor Cinema as a dedicated twin cinema, with two new auditoria, seating 555 and 352, and a shared projection booth. The Village Twin, opened in December 1970, was heralded as "a new concept in cinema entertainment," with the general manager Mr Bruce Simpson adding that it would be "directed at the sophisticated young people of the city ... but would not alienate those of a more conservative outlook."[19] Village opened a series of twins across Australia such as the Double Bay (formally the Vogue cinema) in Sydney in December 1972, followed by the Village Twin in the Sydney suburb of Blacktown, in March 1974. Significantly, later in 1974, Village opened the three-screen Village Cinema City, in the centre of Sydney. This was a purpose-built complex, also designed by Ron G. Monsborough, with auditoria seating 300, 450, and 650 people, with a further two screens added in 1976 and a sixth one in 1986.[20] When Hoyts opened the Chadstone eight-screen complex and precipitated a dash to build multiplexes, all of the major circuits, and smaller, regionally based ones, had recognised the potential of multiscreen cinema. On the eve of the Chadstone multiplex opening, in 1986, there were 723 screens in Australia with three circuits controlling one third of them: Village with 73 screens, Greater Union with 52 screens, and Hoyts with 48 screens.[21]

## THE RISE OF THE 'BIG THREE' MULTIPLEX CIRCUITS

Village Roadshow began, like several US exhibitors, by opening drive-in cinemas as Village-Drive-ins, with its first in Croydon in Melbourne in 1954, before expanding into 'hardtops' in the 1960s and 1970s, becoming Village Theatres. In 1967, it began to vertically integrate its business with the establishment of Roadshow Distributors, which from 1971 saw it handle Warner Bros. films, as well as expanding into TV distribution in 1974 and video in 1985. It was fully vertically integrated with the move into film production in the 1970s, also opening its first theme park—Movie World—in 1991. In the year in which Hoyts opened their

Chadstone multiplex, Village Roadshow opened their four-screen, Village Cinema City, as part of a large real estate development in the centre of Melbourne. Within a year the cinema, which accommodated around 2200 people, had generated some 750,000 admissions, despite being very close to Hoyts' Cinema Centre and Greater Union's Russell Cinema.[22]

Having formed a joint exhibition venture in 1988, with rival exhibitor Greater Union, Village Roadshow opened its first multiplex—the 2500-seat Village Knox 10—two years later, in 1988, in a suburb of Melbourne. It was part of an ambitious building programme that also saw two further ten-screen multiplexes opened in December 1989, in Village Airport West and Village Dandenong, in Victoria. In June of 1989, Warner Bros. International Theaters (WBIT) entered the Australian market by allying with Village Roadshow and Greater Union in a joint venture called the Multiplex Cinemas Partnership, with a professed intention to open a further 200 screens within a few years. WBIT were in the midst of an expansion into several overseas markets, notably in Britain (see Chap. 6).

Greater Union began in 1910 and expanded via a series of mergers, alliances, and joint ventures in the 1920s before the company assumed its current name a decade later. In 1945, the Rank Organisation, Britain's largest film producer and exhibitor, purchased 50 per cent of the company, as the 1950s saw Greater Union ally with rival Hoyts, to open drive-in cinemas and in 1962 it opened a series of 'hardtop' cinemas with other rival Village Theatres (in which it purchased a one third interest). Like Village Roadshow, Greater Union were a vertically integrated company, having formed the distributor British Empire Films (BEF) in 1932, which, with the purchase of MGM's Australian cinema circuit in 1971, saw them gain access to MGM's product and renamed MGM-BEF. They began producing films in 1975, starting with Peter Weir's *Picnic at Hanging Rock* and in 1982 they formed a new distribution company—Greater Union Film Distributors. Upon the purchase of Rank's 50 per cent share of the company, Greater Union was acquired by hotel, leisure, and entertainment group Amalgamated Holdings Ltd in 1984.

Greater Union started building multiscreen cinemas in the 1970s, and, in November 1978, opened a six-screen, 2600-seat complex—the Russell Cinemas—in the centre of Melbourne. The site was that of the former Barclay Cinema and in order to accommodate the six screens the Russell Cinemas had its auditoria on three levels, serviced by three projection boxes. The largest auditorium seated 742 and the smallest 252.[23] At the point in which Greater Union announced its joint venture with Village

Roadshow, it had 63 screens owned outright and 110 in various joint ventures including Village Roadshow and their majority interest in Queensland-based circuit Birch Carroll and Coyle (which they acquired outright in 1991, though the circuit operated under its own brand).[24] As one of three companies in the Multiplex Cinemas Partnership, Greater Union operated a series of multiplexes under its own aegis, of which its first were opened in 1990—an eight-screen site in Hurstville and a nine-screen site in Cambelltown, both in New South Wales, and an eight-screen site in Innaloo, near Perth in Western Australia. When the Innaloo multiplex was expanded to 11 screens in 1994, it was, at that point, the largest multiplex in Australia.[25]

Hoyts was formed in 1926, with the merger of three exhibition companies, though soon after in the 1930s a controlling interest had been sold to the US studio 20th Century Fox, which owned the company until 1982 when it was acquired by a Melbourne-based conglomerate called Stardawn Investments. At that point, the circuit numbered 76 cinemas. In 1985, the Fink family bought out the other partners and formed the Hoyts Corporation in 1986, which set out a four-year building programme with CIC, for the building of 100 screens.[26] Like rivals Village Roadshow and Greater Union, Hoyts were a vertically integrated company having formed their own distribution subsidiary—Hoyts Distribution—in 1978 as well as entering small-scale film production in the 1970s.

Like Village Roadshow, Hoyts too had begun opening multiscreen cinemas in the late 1960s and early 1970s, having formulated a strategy to replace older cinemas in major cities with modern cinemas seating around 1000 people and with the full range of projection technologies (such as 70mm).[27] In Adelaide, the former Regent and Paris Theatres were refitted with smaller auditoria and incorporated shopping arcades. However, Hoyts' key development, predating but clearly inspiring their Entertainment Centre in Sydney, was the three-screen Cinema Centre on Bourke Street in Melbourne, opened in June 1969. The three auditoria were equipped for 70mm projection, with large, curved screens, and eight-track stereo sound, whilst the complex incorporated restaurants and ten further floors of office space. It was widely designated as Australia's first, purpose-built multiscreen cinema and the Southern Hemisphere's largest such cinema at that time.[28] Having opened the Chadstone multiplex in 1986, Hoyts and CIC opened their second site—a 1600-seat, six-screen multiscreen complex—in the Sydney suburb of Pagewood in 1987, followed by an eight-screen multiplex in The Myer Centre in Brisbane, opened in 1988.

## THE MULTIPLEX BOOMS

From the late mid-1980s through to 2000, cinema construction in Australia underwent a boom, as the number of screens tripled in 15 years, though the number of sites underwent little change, meaning that as multiplexes opened, a number of small cinemas closed.[29] In 1986, there were 509 cinemas and 676 screens seating 295,000 people; whilst in 2000, there were 519 cinemas and 1817 screens seating 453,000 people.[30] Most of the construction was multiplexes, with 46 seven- to nine-screen sites opened by 2000 and 35 with ten screens or more, whilst 81 per cent were operated by the three largest exhibitors—Amalgamated Holdings (Greater Union and Birch Carroll and Coyle), Hoyts, and Village Roadshow.[31] Of the multiplexes with ten or more screens in 2000, there were several with 16 screens or more, which meant that they could be classified as megaplexes.

The circuit that was most identified with these larger sites was Greater Union, which opened a 16-screen Castle Hill site near Sydney in 1993, followed in 1994 by the Macquarie 16-screen complex, also near Sydney.[32] However, the largest cinema opened in Australia was Greater Union's 30-screen Megaplex Marion in the Adelaide suburbs of Marion, in November 1997.[33] Part of the Westfield Shopping Centre, the 11,100-square-metre complex could seat 5676 people and was the same size as American Multi-Cinema's (AMC) Ontario Mills complex in California—then the USA's largest cinema (see Chap. 4). The auditoria ranged in size from 547 to 104 seats and had stadium seating, whilst all of the screens were serviced from one 225-metre-long projection room.[34] Serendipitously opening just as big films like *Titanic*, *Tomorrow Never Dies*, and *The Full Monty* were released, the Megaplex Marian was able to run 150 to 180 sessions a day, from 9.30 am to 10.30 pm, utilising 30 to 40 film prints, with a choice of six titles within each hour.[35] Though only Australia's fifth largest city, Adelaide was chosen by Greater Union since a surrounding population of 500,000 people lived within 20 kilometres, and could either drive to the centre and park for free, or use the integrated public transport.

Greater Union were not altogether certain how the megaplex would perform, with Managing Director Robert Manson admitting that the company was "still not sure just what the ideal complex size is [for Australia]. Thirty screens is a stretch, but we have to make it economically viable."[36] In an effort to determine what operating issues might arise, Greater Union sent a delegation to examine similar complexes in Irvine and Dallas in the USA, whilst the experience of partner WBIT's opening of large sites in Britain, including StarCity in Birmingham, was also drawn upon. The Megaplex Marian was indicative of the symbiotic relationship between the multiplex and both the suburbs and the shopping mall. As in the USA, the growth of the shopping mall in Australia took place from the late 1950s onwards, in response to greater suburbanisation, economic growth, the expansion of retailing, and rising car ownership.[37] Unlike in the USA however, Bailey observed, in Australia regional malls tended to be located in established suburban centres rather than on greenfield sites adjacent to freeways, which meant that they had smaller footprints and features such as decked parking.[38]

The attraction of the suburbs and suburban malls for exhibitors can be seen in the changes in the location of cinemas in Australia from the 1980s onwards. In 1985, there were 128 screens located in Australia's five main cities and 167 in suburban areas.[39] By 2000, the comparable totals were 129 in the city and 949 in the suburbs; whilst by 2017 those in the city had dropped to 103, but in the suburbs the number of screens had grown substantially to 1204.[40] As Given, Curtis, and McCutcheon observed, "[f]or those in big cities, 'going to the movies' became a visit to a nearby shopping centre rather than a trip to town. For many in the country towns, it became a trip to a bigger town."[41] Though a significant proportion of Australia's population live in the greater metropolitan areas of its five major cities,[42] there are also a number of large towns and cities, which are defined as "Country' in the official statistics on screens. The numbers of screens there have also risen substantially since the 1980s with 406 in 1985, 700 in 2000, and 903 in 2017.[43] Here, as Aveyard details, the multiplex is also a significant presence, as circuits like Birch Carroll and Coyle, for instance, have opened multiplexes in Cairns and in Coffs Harbour.[44] Though owned by Amalgamated Holdings Ltd, Birch Carroll and Coyle have been a significant multiplex operator, with ten multiplexes of seven to nine screens and seven with ten screens or more in 2017, substantively located in Queensland, where the company is based.[45]

Though the exhibition market in Australia is dominated by the 'Big Three' of Amalgamated Holdings, Hoyts, and Village Roadshow, and has been throughout the multiplex era, this is not to suggest that there have not been some substantial changes in the complexion of the market. Unlike many other countries in this study, exhibitors in Australia have displayed a consistent willingness to cooperate via the creation of alliances and joint ventures. Though Village Roadshow and Greater Union bought out WBIT's share in the Multiplex Cinemas Partnership in 2003, the two companies have continued their joint venture as Australian Theatres. In 2009, Amalgamated Holdings launched a new brand called Event Cinemas, and this has been adopted in many of their new and renovated multiplexes as part of the Australian Theatres partnership.[46] In 2015, Amalgamated Holdings Ltd changed its name to Event Hospitality and Management Ltd and, as of 2017, they were the largest exhibitor and multiplex operator in Australia, with 63 sites of four screens or more, followed by Hoyts with 38 and Village Roadshow with 21.[47]

## DIVERSIFICATION—NEW ENTRANTS TO THE MULTIPLEX MARKET

Though the position of the three largest exhibitors is fairly assured there has been a significant development in the multiplex market in Australia, as Given, Curtis, and McCutcheon detail, which is the entry of two national arthouse circuits: Palace and Dendy.[48] Palace, which has been a distributor since 1965, is also the largest arthouse circuit in Australia opening cinemas in Melbourne in the 1970s before expanding through both acquisition and construction.[49] Like other exhibitors, it was not averse to partnerships, signing a 50:50 joint venture with Village Roadshow in 1994, and further ventures with Adelaide-based Nova Cinemas in 1999 and Perth-based Luna Cinemas in 2000. By 2007, when Palace owner Antonio Zeccola bought out Village Roadshow's share, it operated ten multiscreen cinemas including the Adelaide Nova, a seven-screen, 1400-seat multiplex.[50] In 2017, it was Australia's fifth largest exhibitor with 124 screens in 24 sites, and operated seven multiplexes with 7 or more screens, including the 12-screen Palace Raine Square located in a shopping complex in Perth.

Dendy Theatres, a subsidiary of distributor Filmways, opened a series of twin cinemas in its home city of Melbourne in the late 1970s, as well as in other locations such as Sydney's Martin Place, opened in 1981 and closed in 2003, whereupon Dendy operated five multiscreen sites with 16 screens.[51] As an important distributor as well as an exhibitor, Dendy was purchased by the media company the Becker Group in 1997 and in 2008 it became the Dendy Icon Group, having been purchased by Mel Gibson and his partner in Icon Productions, Bruce Davey. It has become more ambitious in its building programme, opening three multiplexes including the 13-screen Dendy Canberra and two 10-screen sites in Coorparoo in Queensland and Newtown in New South Wales.

Though no US operators made inroads into the Australian exhibition sector in the first ten years of the multiplex's development, apart from those in alliances such as Warner Bros., in 1996, US company Reading opened its first multiplex, a six-screen site in Townsville, Queensland, followed by a second six-screen site in Mandurah, Western Australia, in 1997. Reading was part of Reading International, an investments company, which operated cinemas in the USA and Puerto Rico. It entered the Australian market with bold ambitions to open 150 screens in five years, including an announcement in December 1995 that it had acquired land in Burwood, a suburb of Melbourne, for a 25-screen megaplex, which would have then been the largest cinema in Australia.[52] Reading's strategy was to purchase their own land for construction, rather than to open in shopping developments with the associated issues with leaseholds, though the proposed Melbourne megaplex was rejected by the Victoria state government after vigorous lobbying from Hoyts and Village Roadshow.[53]

Reading's entry into the Australian market ran up against considerable, organised resistance from the three established domestic exhibitors. In Penrith, in the suburbs of Sydney, Reading's plans for a 12-screen multiplex was opposed by both the local independent exhibitor, Hayden Cinemas, which operated Penrith's only cinema, and Hoyts, which had council approval for a 10-screen cinema in the area.[54] Many of Reading's subsequent schemes faced opposition from the major exhibitors, such as in Liverpool, near Sydney, where Readings proposed a 20-screen complex less than a mile from Greater Union's expanding 12-screen site.[55] In the case of both Penrith and Liverpool, Reading insisted that the areas were under-screened. It was not just through the planning process that Reading ran into opposition from the incumbent exhibitors, since they were in a position to influence the major distributors in restricting access to films. When Reading opened their Market City 5 cinema in the centre of Sydney, they were denied access to first-run films by the major distributors, one of which was Village Roadshow, because of the cinema's proximity to those of Hoyts, Village Roadshow, and Greater Union.[56]

The problems faced by Readings were seen as clear evidence of what many had observed for several years, and that was that the exhibition industry in Australia was too concentrated and anti-competitive. As Jones observed, independents had long:

claimed that the major distributors, particularly Roadshow and UIP, have a substantial degree of market power ... [s]pecifically it is argued that Roadshow uses its market power to damage independent exhibitors, to the benefit of Roadshow's exhibition partners Greater Union/Village/Warner Bros or some combination thereof.[57]

As a result in part of Jones' recommendations, in 1998 the Australian Competition and Consumer Commission launched a voluntary Code of Conduct for Film Distribution and Exhibition, which covered, amongst other things, supply of film prints and a mechanism to resolve conflicts.[58] For several years, Reading Cinemas' Sydney site operated as a second-run cinema before Reading appealed to the new Australian exhibition industry's code-of-conduct committee, from which it won a favourable ruling in 2000.[59] For Reading Cinemas' CEO John Rochester, Australia proved to be "one of the most aggressive markets in the world."[60] Notwithstanding their travails, Reading is now the fourth largest circuit in Australia with 149 screens.[61]

## Exporting the Multiplex

The aggressive nature of the domestic market was mirrored by the aggressive expansionism of the three major exhibitors, all of whom looked abroad in order to find new markets. Village Roadshow entered into a series of joint ventures with exhibitors in several countries in both Asia and Europe. Having already established a partnership in Australia, Village Roadshow and WBIT formed Warner Village Cinemas Ltd in 1996, in order to manage their circuits in Britain, which prior to the sell-off in 2003 and retreat from the British market numbered 36 multiplexes, including a 30-screen megaplex in Birmingham's StarCity (see Chap. 6). Village Cinemas International, the company through which Warner Village Cinemas ran its European operations, had a presence in several markets including Austria, Germany, Greece, Italy, and Switzerland, though by the early 2000s, when it had more than 115 screens, it was in the process of exiting many of these markets and selling out to domestic exhibitors.[62] In Asia, Village Roadshow similarly formed a series of joint ventures, notably with Indian exhibitor Priya, with whom they opened India's first purpose-built, multiscreen cinema in Anupam in 1996, as well as with Tanjong in Malaysia, Cheil Jedany in South Korea (see Chap. 11), and Entertain Theatres Network in Thailand. Having formed Golden Village with Hong Kong's Golden

Harvest, they opened what was claimed as "Asia's first multiplex"—the Yishun ten-screen in Singapore—in 1992. However, by 2017 and the selling off of its share in Golden Village, the company had retrenched back to Australia, with just a remaining stake in the US iPic Theatres circuit.[63]

Early on, Hoyts looked to expand overseas and, in September 1986, it acquired Cinema Centers Corp., based in Boston, which had opened the influential Crossgates Mall in Albany in 1984 (see Chap. 3). In addition, it also purchased another US circuit, SBC Theaters, in December 1986, which, along with a series of other smaller acquisitions in the following three years, meant that in 1988, the company was the ninth largest circuit in the USA.[64] By 1999, it operated 116 sites with 945 screens in the USA, as well as operations in Mexico, with a 36 per cent stake in Cinemex, and a joint venture with General Cinemas in Chile. Hoyts had sold out its US and Mexican operations to a range of rivals, however, by 2003. Hoyts' other significant overseas venture was a joint undertaking with Thiele Entertainment in Germany, began in 1998, though just two years later Hoyts sold its interest in the group's ten multiplexes to domestic operator Kinowelt. Having expanded and contracted Hoyts was itself acquired by AMC, owned in turn by Chinese company Wanda Cinema Line Corp., part of the Dalian Wanda Group, in 2015, for an estimated A$900 million (US$750 million).[65]

Greater Union expanded into overseas markets, notably in the United Arab Emirates, where it opened a six-screen multiplex in Dubai in 1999, in partnership with shopping mall developer Majid al-Futtaim, as well as the Netherlands and Poland. However, its only presence outside of Australasia now is in Germany, having purchased 50 per cent of domestic exhibitor Kieft & Kieft in 1998. Unlike Hoyts, Greater Union (now Event) has stayed the course in Germany largely as a result of having to buy out its former partner in 2003 in order to avoid them going bankrupt. Nevertheless, the newly named Event/CineStar chain is the largest in Germany with 426 screens (see Chap. 9).

Despite retrenchment back into their domestic markets, the major circuits, along with Palace, Dendy, and Reading, have not stopped building multiplexes in Australia. Though the boom, which saw many multiplexes built in the wake of the first multiplex in 1986, had tapered off by 2000, there is still room for development. Though resolutely a suburban phenomenon, there are strong indications that the city centre is becoming the focus for development, with a 27 per cent increase in the number of cinemas opened in central business districts between 2016 and 2017.[66]

# NOTES

1. Jock Given, "The Multiplex Era," in *American–Australian Cinema: Transnational Connections*, eds. Adrian Danks, Stephen Gaunson and Peter C. Kunze (London: Palgrave Macmillan, 2018).

2. Screen Australia, *Cinema Industry Trends: Historical Overview of Admissions Since 1901*. https://www.screenaustralia.gov.au/fact-finders/cinema/industry-trends/historical-admissions (accessed 10 January, 2019).

3. Ibid.

4. Ibid.

5. Ibid.

6. Screen counts are taken from: *Film Weekly: Motion Picture Directory 1950–51*: 72–96; *Film Weekly: Motion Picture Directory 1960–61: 40–60;* and *Film Weekly: Motion Picture Directory 1971*: 20–27 (Sydney: Film Weekly PTY. Ltd).

7. Dodona, *Cinemagoing Asia Pacific 2004* (Leicester: Dodona Research, 2004), 24; and *Cinema Industry Research Australia & New Zealand 2018* (Leicester: Dodona Research, 2018), 3.

8. See Given, "The Multiplex Era."

9. The multiplex was expanded to a 17-screen megaplex in 1999 but closed in 2014 and demolished. Hoyts opened a new 13-screen multiplex on the site in 2016.

10. Quoted in Andrew L. Urban, "Cracking the Nut," *Screen International* 734 (9 December, 1989), 24.

11. Ibid., 24.

12. "Filmgoers Flock To Oz' First Multiplex," *Variety* 324, no. 10 (31 December, 1986), 23.

13. Blake Murdoch, "Hoyts, CIC Buddy Up To Spread Multiplexes Through The Suburbs," *Variety* 324, no. 3 (13 August, 1986), 61.

14. Ibid.

15. The complex opened officially on 7 January 1977 with US film star Gregory Peck in attendance. For a consideration of the complex's projection facilities and films shown, see Peter Fraser, *Sydney in 70mm—Part Three The Multiplex: Hoyts Entertainment Centre*, https://www.in70mm.com/news/2016/louden/index.htm (accessed 20 January, 2019).

16. Hoyts Theatres, Advertisement for Entertainments Centre, *Variety* 291, no. 1 (10 May, 1978), 49–52.

17. "6 Reasons Why Hoyts Australia is One of the Top Grossing Cinema Chains in the World!" *Variety* 282, no. 12 (5 May, 1976), 71.

18. "Hoyts Chain in Turnaround Via Lift from Sydney Septplex," *Variety* 291, no. 1 (10 May, 1978), 53.

19. Queensland Government, *Queensland Heritage Register: Village Twin Cinemas* (20 January, 2016). https://apps.des.qld.gov.au/heritage-register/detail/?id=602101 (accessed 20 January, 2019).

20. It closed in 2000 and was subsequently demolished to make way for apartments.

21. "Aussie Screens," *Variety* 323, no. 1 (30 April, 1986), 61.

22. "Village Roadshow Finds Middle Titles also Getting Results," *Variety* 327, no. 1 (29 April, 1987), 58.

23. It was closed in October 2013 and demolished a year later to make way for a hotel.

24. "Greater Union Cinema Div. Back to High Profile Status," *Variety* 330, no. 13 (20 April, 1988), 39.

25. In 1998, it was refurbished and re-opened with 16 screens. See Western Australia Cinemaweb. http://www.ammpt.asn.au/CinemaWEB/SITE/view.php?rec_id=0000000404 (accessed 28 January, 2019).

26. In 1989, Hoyts' CEO Peter Ivany acknowledged that this target was "slightly optimistic" and was 6–12 months behind schedule. See *Variety*, "Hoyts Shoring Up Firm, Cuts Debt Before Next Round of Expansion," 335, no. 1 (26 April, 1989), 124.

27. Dale Turnbull, "Australia Rebuilding; Theatres In Plazas Share Lobby Like In U.S.," *Variety* 253, no. 8 (8 January, 1969), 129.

28. See Dale Turnbull, "Confidence: It Builds Theatres," *Variety*, 257, no. 8 (7 January, 1970), 6.

29. Dodona, *Cinemagoing Australasia 2006* (Leicester: Dodona Research, 2006), 6.

30. Ibid., 7.

31. Screen Australia, *Cinema Industry Trends: Multiplex Sites by Exhibitor.* https://www.screenaustralia.gov.au/fact-finders/cinema/industry-trends/screens-and-theatres/multiplexes-by-exhibitor (accessed 19 January, 2019).

32. Both cinemas were subsequently reduced in size, with Castle Hill becoming ten screens in 1999 and Macquarie eight screens in 2000.

33. The complex was run by Greater Union but was built as part of the Multiplex Cinemas Partnership, with Village Roadshow and WBIT.

34. Sandy George, "Greater Union's Mega Hit: 30-Screen Complex Woos Moviegoers in Adelaide Suburb of Marion," *Film Journal International* 101, no. 3 (1 March, 1998), 80, 253.

35. Ibid., 80.

36. Quoted in Blake Murdoch, "A Greater Union: Leading Aussie Circuit Plans to Double Its Screen Count," *Film Journal International* 100, no. 1 (1 February, 1997), 100.

37. Matthew Bailey, "Shopping for Entertainment: Malls and Multiplexes in Sydney, Australia," *Urban History* 42, no. 2 (2014): 309–29.

38. Ibid., 312.
39. Screen Australia, *Cinema Industry Trends: Screens by Type of Location, 1985–2017*. https://www.screenaustralia.gov.au/fact-finders/cinema/industry-trends/screens-and-theatres/type-of-location (accessed 19 January, 2019). 'City' refers to the central business districts (CBD) of Sydney, Melbourne, Brisbane, Adelaide, and Perth; and 'Suburban' refers to the cinemas in non-CBD areas of those five cities.
40. Ibid.
41. Jock Given, Rosemary Curtis and Marion McCutcheon, *Cinema in Australia: An Industry Profile* (Melbourne: Swinburne Institute of Technology, 2013), 11. http://apo.org.au/system/files/34831/apo-nid34831-88216.pdf (accessed 24 January, 2019).
42. According to the Australian Bureau of Statistics, of Australia's 24.6 million population in 2017, 65 per cent lived in Greater Sydney, Melbourne, Brisbane, Adelaide, and Perth. http://www.abs.gov.au/ausstats/abs@.nsf/Latestproducts/3218.0Main%20Features12016-17?opendocument&tabname=Summary&prodno=3218.0&issue=2016-17&num=&view (accessed 26 January, 2019).
43. Screen Australia, *Cinema Industry Trends: Screens by Type of Location*.
44. Karina Aveyard, *Lure of the Big Screen: Cinema in Rural Australia and the United Kingdom* (Bristol: Intellect, 2015), 22–3.
45. Screen Australia, *Cinema Industry Trends: Screens by Type of Location*.
46. See Given, Curtis and McCutcheon, *Cinema in Australia*, 21.
47. Screen Australia, *Cinema Industry Trends: Multiplex Sites by Exhibitor*.
48. Given, Curtis and McCutcheon, *Cinema in Australia*, 24.
49. See Michael Kitson, "The Rise of the Boutique Cinemas: New Trends in Cinema Exhibition," *Cinema Papers* 107 (December, 1995): 14–17.
50. Screen Australia, *Cinema Industry Trends: Multiplex Sites by Exhibitor*.
51. Dodona, *Cinemagoing Asia Pacific 2004*, 49.
52. "Reading Enters Oz," *The Film Journal* 99, no. 2 (1 February 1996), 94.
53. George Sandy, "Readings in Australia," *Film Journal International*, 101, no. 4 (1 April, 1998), 45.
54. Don Groves, "Oz Exhibs Battle Rookie in Suburban Screen Skirmish," *Variety* 365, no. 1 (4 November, 1996), 19.
55. Ibid.
56. Francesca Dinglasan, "Justice for Reading?," *BoxOffice* 136, no. 1 (1 November, 2000), 150.
57. Ross Jones, *Developments in the Cinema Distribution and Exhibition Industry: Report to the Australian Competition and Consumer Commission* (March, 1998), 48. https://www.accc.gov.au/system/files/The%20Cinema%20Industry.pdf (accessed 22 January, 2019).

58. Australian Competition and Consumer Commission, *Film Distribution and Exhibition Code of Conduct*, 13 August, 1998. https://www.accc.gov.au/media-release/film-distribution-and-exhibition-code-of-conduct (accessed 26 January, 2019).
59. Francesca Dinglasan, "Reading And No Longer Waiting For First-Run Films," *BoxOffice* 136, no. 12 (1 December, 2000), 50.
60. Susan Lambert, "The Upside From Down Under: A Look At Australia's Booming Exhibition Industry," *BoxOffice* 133, no. 8 (1 August, 1997), 28.
61. Dodona, *Cinema Industry Research Australia and New Zealand 2018*, 4.
62. Dodona, *Cinemagoing Western Europe 2001* (Leicester: Dodona Research, 2001), 60.
63. Dodona, *Cinema Industry Research Australia and New Zealand 2018*, 4.
64. "Ten Giants: Hoyts," *BoxOffice* 124, no. 12 (1 December, 1988), 26.
65. Patrick Frater, "China's Wanda Buys Australia's Hoyts Multiplex Chain," *Variety* (2 June, 2015) https://variety.com/2015/film/asia/chinas-wanda-buys-australias-hoyts-1201510298/ (accessed 23 January, 2019).
66. Dodona, *Cinema Industry Research Australia and New Zealand 2018*, 3.

# CHAPTER 11

# The Multiplex in Japan, South Korea, and China

The Asia-Pacific region covers an enormous area geographically and culturally, and its cinema markets are far from homogenous. Nevertheless, whilst the patterns of development within those territories are not consistent, the diffusion of the multiplex cinema has penetrated all cinema markets. The region was one that US exhibitors began to look at in the early 1990s. For American Multi-Cinema (AMC), anxious to expand in the mid-1990s, they saw "tremendous opportunity in the Pacific Rim," adding that "most Asian nations are greatly underscreened, and all the major countries are western-looking, active consumers of entertainment."[1] In terms of the international box-office, Asia-Pacific has been the fastest growing region with the 2018 total reaching $16.7 billion, a 35 per cent rise in the period since 2014.[2] The increase in box-office reflects, in part, the increase in admission prices in all countries as the ageing cinema infrastructure has been replaced with modern multiplex cinemas. As Dodona observed: "[i]n many ways what might be termed the multiplex revolution is really the replacement of a cheap, not necessarily high quality form of mass entertainment with a more costly, more luxurious and more restricted entertainment form."[3]

In terms of international box-office markets in the region, the two largest are China and Japan, with South Korea fourth.[4] Whilst China dominates in terms of numbers of screens and box-office, this is a relatively recent development, especially since the numbers of screens in China was only around 2200 in 2000, compared to 2544 in Japan and 720 in South

© The Author(s) 2019
S. Hanson, *Screening the World*,
https://doi.org/10.1007/978-3-030-18995-2_11

Korea.[5] By 2017, all markets had expanded with South Korea almost quadrupling to 2766 screens,[6] and Japan more than doubling to 3525.[7] However, growth in China had exploded with 50,776 screens in 2017.[8] Again, in part this reflects the relative maturity of different markets in the region, with both Japan and South Korea having more established multiplex infrastructures.

Therefore, this section considers three national case studies—Japan, South Korea, and China—since all highlight different trajectories in the development of the multiplex. Japan was for a long time the second largest cinema market in the world and is one in which the multiplex has grown rapidly. However, unlike other countries in this chapter, it experienced significant penetration by US exhibitors in the early days of the multiplex's development. South Korea is now the highest-attending cinema market in the world, whilst the number of multiplexes has grown significantly since the late 1990s.[9] Its exhibition sector is dominated by domestic operators, which have expanded into neighbouring countries. Finally, usurping Japan as the second largest cinema market in the world in 2012, China's exhibition sector has undergone incredible expansion, though its rate of cinema attendance per capita is extremely low by any comparison. Its emergent cinema industry, closed tied to the importance of soft power and the projection of Chinese values, is reflected in the global expansion of some of its companies, notably the Dalian Wanda Group. Wang Jianlin's company, through its subsidiary Wanda Cinema Line, is the largest exhibitor in the world.

## Japan

Japan's first multiplex cinema was opened in Ebina, in April 1993, by Warner-Mycal; a 50:50 joint venture between US company Warner Bros. International Theaters (WBIT) and Japanese supermarket and retail group Nichii Co. The seven-screen complex, with 1970 seats, was located in a shopping complex owned by Nichii, in Kanagawa, within the greater-Tokyo metropolitan area. This first multiplex, followed in the same year by both an eight-screen complex in Kishiwada near Osaka and a six-screen one in Takaoka near Toyama, was part of a plan to open 30 multiplexes in the following six years, with Nichii building the cinemas and WBIT operating them.[10] The advantages of the joint venture for WBIT was that with the sites owned by Nichii, they had plenty of car parking and WBIT did not have to purchase some of the most expensive land in the world.

Moreover, Warner-Mycal's strategy was to open their cinemas in satellite towns around big cities that had been vacated by the major domestic distributors/exhibitors—Shochiku, Toho, and Toei—which owned some 90 per cent of Japan's cinemas.[11] Warner-Mycal proposed to challenge the domestic exhibitors' lack of investment and innovation in their cinemas, especially in the context of declining admissions, and the realisation that some 60 cities with 100,000 inhabitants or more had no cinema.[12] Satoru Koyama, from the Ministry of International Trade and Industry, observed that "[p]eople want to see Japanese movies, but they don't want to go to the movie theaters."[13]

Prior to the opening of the Warner-Mycal Ebina, the cinema infrastructure in Japan had, like all the countries in this study, contracted since the 1950s and/or 1960s. In 1960, at its peak, there were 7457 screens in Japan, whilst the cinemagoing audience had peaked in 1958, with 1.12 billion visits.[14] Thereafter, the number of screens declined until it reached 1734 in 1993, whilst admissions had bottomed out at 119.5 million in 1996.[15] As in Britain, for instance, the entry of a US exhibitor into the market had been just as cinemagoing and the cinema infrastructure were experiencing a dramatic and accelerated decline. Moreover, as in Britain, the domestic exhibitors, which dominated the market, were, in the words of Dodona, "taken rather unawares by the emergence of a multiplex."[16] In the decade after the opening of the first multiplex, the boom in building saw a 35 per cent increase in the number of screens,[17] with the increase accounted for by those showing both Japanese and foreign-language films.[18] This reflected the number of multiplexes being opened, which by 2008 contained 2659 screens (79 per cent of the total),[19] and 3096 screens by 2017.[20] Warner-Mycal's strategy to boost admissions included offering reduced-price tickets for later performances, as it was observed that traditionally most cinemas in Japan put on their last showings at 7 pm due to reluctance on the part of audiences to go out later.[21] Moreover, given Japan's importance as a market for US films, and its domestic film production, Warner-Mycal's programming strategy was for a mix of both, from a range of distributors.[22]

Warner-Mycal were the most prolific multiplex builders during this period, having opened an impressive 44 multiplexes containing 337 screens in their first ten years of operation, which averaged over 4 sites a year and 14 in 2000 alone.[23] WBIT's entry into the Japanese market was the first of several for overseas exhibitors, most notably American Multi-Cinema (AMC), which opened Japan's then-largest multiplex cinema—

the AMC Canal City 13—in Fukuoka, in April 1996. Unlike Warner-Mycal's multiplexes, built largely on greenfield sites, in smaller cities, AMC opted to incorporate their multiplex into a giant new downtown hotel and retail complex—Canal City—in Japan's eighth largest city. AMC's Chip Harris felt that Japan was a "very fertile ground" for the company, adding that it "presents us with a highly sophisticated audience" that is "very Western looking and forward-looking in entertainment."[24]

AMC's entry into the market meant that along with Warner-Mycal, access to first-run films was hampered by the control exerted by the major domestic distributors, which were also the major exhibitors. At Fukuoka, where AMC opened in a location that had existing cinemas, their efforts to programme the major releases on-date were often delayed by distributors anxious to advantage their own cinemas. In the case of *Princess Mononoke*, which took over $150 million at the Japanese box-office in 1997, the film's distributor Toho offered the film to AMC only two weeks after its release.[25] Nonetheless, AMC claimed to have generated one million admissions in the first year of operation and 1.5 million in the second,[26] which meant that it captured more than 50 per cent of the market and increased cinemagoing in Fukuoka by 30 per cent in the first year.[27]

AMC's targeting of Japan was part of a broader strategy of expansion in Asia, including Hong Kong, where they opened an 11-screen multiplex in December 1998, though the focus was on Japan as the largest market. Of the first five multiplexes AMC built in Japan, four were megaplexes and the largest in the country, including 16-screen sites in Nakama, Tokyo, and Shinsei, opened between the end of 1998 and the middle of 2000, and the 18-screen Holiday Square complex in Toyohashi. Like Warner-Mycal, AMC sought to stretch out the programming into the later evening and instigated screenings at 9–10 pm, which proved to be popular, accounting for 25 per cent of individual cinema's business.[28]

United Cinemas International (UCI), co-owned by Universal Studios and Paramount Communications, and one of the largest exhibitors in Britain and Europe (see Chaps. 6 and 9), entered the Japanese market in November 1996, opening the 1816-seat Otsu 7, near Kyoto. UCI's first foray into Japan led them to open a smaller multiplex on the top two floors of a department store, on the basis that though Warner-Mycal and AMC had paved the way, the average Japanese cinemagoer was "conservative by nature" and that traditionally, they were "not used to multiplexes, and there's an attitude that they don't provide good value for money."[29] UCI's second site—the 2276-seat, Renais 9—opened in Kanazawa in

1997—was its first freestanding multiplex, though located within the Renais entertainment resort, with its attendant free parking for 2000 cars. Like its foreign predecessors, UCI opted to open until late at night, and programmed both Japanese and foreign-language films.

British exhibitor Virgin Cinemas established its Japanese operation in 1999 and opened three multiplexes in its first year: a 14-screen site in Hisayama; a 12-screen site in Nagoya and a 9-screen site on Ichikawa, with a proposal to open 20 multiplexes within five years. Virgin had a strong brand presence in Japan, having opened a series of its Megastores, beginning with one in Tokyo's Shinjuku district in September 1990. In the event, Virgin opened eight multiplexes, with 81 screens before selling out to Japan's largest exhibitor Toho in February 2003. At this point, Virgin's exit from the Japanese market was mirrored by several other overseas exhibitors.

By 2000, UCI had opened ten multiplexes with a total of 89 screens, though only the first three opened were operated solely by UCI, since in 1999 they entered into a joint venture with the media company Sumitomo Corporation and film producer Kadokawa Shoten. With UCI owning 50 per cent of the company, the newly named United Cinemas Japan planned to open five further multiplexes. However, in 2004, UCI sold out its 50 per cent stake to Sumitomo, which meant that United Cinemas, with 89 screens, had become both a domestically owned circuit and the fourth largest in Japan.[30] In 2005, United Cinemas expanded after acquiring AMC's multiplexes, as that US company exited the Japanese market. This meant that after disposing of one of those sites, it retained 15 multiplexes and 161 screens, making United Cinemas the third largest circuit. This left WBIT as the only foreign multiplex operator with a presence in the Japanese market.

It is clear that US-based exhibitors kick-started the development of the multiplex in Japan, joined by those from Britain. They did so in the context of an exhibition sector dominated by three large domestic, vertically integrated companies, all of which had concentrated on a pre-existing infrastructure of single-screen and small 'miniplexes,' along with a strict demarcation between cinemas that showed Japanese films and those which showed foreign-language films. All saw their market shares fall as the overseas companies opened multiplexes. In 1997, Toru Okuyama, CEO of Shochiku, the oldest of the three major film companies and exhibitors, outlined the exhibition landscape in Japan as: "characterized by high admissions price, lack of screens, and an old exhibition system. Block booking and track booking still exist here."[31]

When Warner-Mycal opened Japan's first multiplex, the country's largest exhibitor was Toho, which operated traditional cinemas, many of which had been converted into two- and three-screen venues, whilst rather than build multiplexes as a way of competing with the new market entrants, it chose to use its power as a distributor to restrict their access to films, especially where they went up against their own sites. When it did enter the multiplex market substantively in 2003, it did so via the acquisition of Virgin Cinemas' eight multiplexes. The Virgin sites complemented Toho's existing small number of multiplexes, including its first foray into multiplex building with a five-screen site in Otsu and a six-screen site in Fukuoka, both opened in 1997, and a seven-screen site in Kagashima in 1998. The most significant development, however, was the 13-screen, 3000-seat Toho Cinema Mediage, located in the Sony Entertainment Centre in central Tokyo, which opened in 2000. It was the first multiplex opened in the centre of the capital city. By March 2018, Toho was the second largest exhibitor with 66 sites and 772 screens.[32]

Shochiku, which, like Toho, was a vertically integrated company, was the first of the major domestic circuits to build multiplex cinemas when its new subsidiary Shochiku Multiplex Theatres (SMT) formed an alliance with US exhibitor Cinemark to open a number of sites under the brand name Movix. The first multiplex to open was a seven-screen multiplex in March 1997, on the ninth floor of the Rink Kobe Fashion Plaza, a shopping complex on Rokko Island. Despite setting out plans to open ten multiplexes by 2000, the joint venture with Cinemark lasted less than a year, as in the face of losses, Shochiku dissolved it. Nonetheless, SMT set in train a series of developments which saw it operating 11 multiplexes by 2003, with 99 screens under the Movix brand, in addition to 59 screens in its traditional sites.[33] By March 2018, Shochiku was largely a multiplex operator, having greatly slimmed down its traditional sites, whilst its 22 sites and 227 screens meant that it was now the fourth biggest circuit.

Toei was the third of the original major domestic exhibitors, and was the last into multiplex construction, forming a subsidiary called T-Joy in 2000, with plans to open four multiplexes within four years. It opened its first—the six-screen T-Joy Higashi in Hiroshima—in 2000 and installed digital projectors that could take digitised films downloaded from a satellite feed.[34] By the end of 2003, T-Joy had opened six multiplexes with 47 screens, all of which had some of the auditoria equipped with digital projectors. As a vertically integrated company, Toei's production arm was also pioneering digital filmmaking. T-Joy, in partnership with rival Toho, operates Japan's largest cinema—the Wald 9 in Shinjuku, Tokyo—located on the 9th to 14th floors of the Shinjuku 3-Chrome East Building, and opened in 2007. A further Wald cinema with 11 screens was opened in Hiroshima's Aeon Mall in 2007. The name Wald derives from the German for forest, since the screens are so large they surround the cinemagoer like a forest.[35] By March 2018, T-Joy was the fifth biggest exhibitor, with 14 sites and 131 screens.

By 2012, Warner-Mycal were the second largest circuit behind Toho with 60 sites and 496 screens, though they had by far the largest number of multiplex screens and the only one with any foreign interest.[36] However, in the same year WBIT's interests in the joint venture were bought by Aeon Cinemas, which up until then had been a small chain of 13 multiplexes with 103 screens.[37] Aeon was a large retail group, which had purchased the Mycal group in 2003 and had thus become the other half of the Warner-Mycal joint venture. The merging of the former Warner-Mycal and Aeon Cinemas meant that Aeon became the largest exhibitor by screens in Japan, and in March 2018 it operated 91 sites with 772 screens.[38] Japan's exhibition market is now mature and wholly domestically owned, with attendances still high. Indeed, in 2016 and 2017 admissions were extremely healthy, as more Japanese people went to the cinema than at any point since the mid-1970s.[39] However, unlike their near neighbours in South Korea, for instance, Japanese exhibitors have no presence in any other country, preferring to focus on domestic markets.

## South Korea

South Korea's exhibition sector has a high proportion of multiplexes—361 out of 452 cinemas in 2017, which is 79.9 per cent of the total.[40] In terms of exhibitors, South Korea is a highly concentrated market, with three circuits—CGV, Lotte Cinema, and Megabox—accounting for 354 cinemas, which is 78.3 per cent of the total.[41] When one considers multiplexes, then their domination is complete since their sites contain 92 per cent of the country's 2766 screens.[42] CGV, Lotte Cinema, and Megabox have all sought to expand overseas, with near neighbours Vietnam and China the primary targets.

South Korea's pattern of cinema attendance has been quite unlike other countries in this study, largely because of the complex socio-political history of the country. In terms of trends, South Korea's cinema audience increased in the period after the end of the Korean War armistice and through the 1960s, with 98.3 million admissions in 1960, rising to a peak of 168.4 million in 1970, before a precipitous fall throughout the 1970s until 1986 when it had halved to 47.3 million.[43] In terms of cinemas, before the outbreak of the Korean War in 1950, there were 116, which had increased to 658 by 1970, before declining to 447 in 1985.[44] The mid-1980s were the low point in both the cinema infrastructure and admissions, as the declines since 1960 coincided with Park Chung-Lee's authoritarian government, which passed the Motion Picture Law of 1962, limiting the numbers of films both produced and imported, as well as imposing strict, ideological censorship.[45] Cinema audiences declined as a result and it was only in the late 1980s, with the revision of the Motion Picture Law, the loosening of censorship and limits to production and distribution, and the permitting of Hollywood studios to open branch offices, that admissions began to stabilise and then rise.[46] The number of cinemas also began to rapidly increase after 1985 too, with 507 in 1998, though given that the population was some 146 million, this meant that South Korea was seen by many as under-screened.[47]

The year 1998 was significant, since in April the first multiplex in South Korea—the CGV Riverside 11—opened in the outskirts of Seoul. The multiplex was located on the tenth floor of the Technomart 21 complex, which had 45 storeys, a settled population of 18,000, and a commuting population of 40,000.[48] CGV was a joint venture between Cheil Jedang, a food and chemical company via its entertainment division CJ Entertainment, Hong Kong–based Golden Harvest, and the Australian exhibitor Village Roadshow.[49] In 2000, Village Roadshow bought Golden Harvest's share, though in 2002 it sold out its holdings as it exited the Korean market. CGV opened a further multiplex in December 1999—the CGV 14 in the port city of Incheon—before accelerating construction and opening 4 sites in 2000 and a further 11 by the end of 2003, including 4 in South Korea's second largest city, Busan, and a further 4 in Incheon.[50] In all instances, the multiplexes were located in larger shopping and leisure complexes, in part because of the high land costs allied to urban density. In the case of the sites in Seoul, like the Riverside 11, CGV looked for easy access for up to one million people, with public transport links extremely important and ever conscious of choosing the right location due to the minimum leasing contract of 20 years.[51]

CGV's expansion continued throughout the 2000s and in 2006, they formed an agreement with Korean studio Prime Entertainment to operate a series of sites branded initially as CJ Prime and latterly as Primus. More significantly, perhaps, have been its forays into neighbouring countries, with a joint venture called CJ Century with Chinese company Shanghai Film Group, which opened a six-screen multiplex in Shanghai's Daning Commercial and Cultural centre, in October 2006. In July 2011, CGV entered the Vietnamese market by purchasing Megastar Cineplex's seven multiplexes with 54 screens, which in 2017 had grown to 50 sites and 324 screens.[52] By August 2018, CGV was operating in several other world markets including Turkey, Indonesia, and Hong Kong, in addition to major expansion in China, with 304 cinema sites abroad.[53] As the dominant exhibitor in South Korea, CGV has 145 sites in South Korea with 1085 screens.[54]

Lotte Cinema, established in 1999, is South Korea's second major exhibitor, with 114 sites and 810 screens,[55] including South Korea's largest cinema—the 21-screen, 4165-seat Lotte Cinema World Tower, opened in 2014, in the Lotte World mall in Seoul. Lotte was a large retail conglomerate and department store chain, which began opening multiplexes in its own shopping malls, with its first a six-screen site in Ilsan, northwest of Seoul, in October 1999. Lotte then opened further sites the following year, in Daejeon and Gwangji, on the upper floors of its shopping developments.

Like CGV, Lotte has sought to expand overseas and, as with CGV, has looked to both Vietnam and China. In the case of the former, Lotte purchased Vietnamese circuit Diamond Cinema Joint Venture Company in 2008, and opened multiplexes in Vietnam's capital Ho Chi Minh City and other large cities. By 2018, it had 18.2 per cent of the market and 40 sites with 150 screens, which, along with CGV, meant that the South Korean companies controlled over 63 per cent of the Vietnamese market.[56] Insofar as expansion into mainland China is concerned, Lotte, like CGV, purchased a single-screen site in Hong Kong in order to facilitate this administratively, forming a joint venture called Lotte & NN Cinema and opening a six-screen multiplex in Shenyang in 2010. By 2018, Lotte Cinema China had 11 multiplexes in China with 87 screens, when Lotte had also restructured its business, in order to spin off its film production, distribution, and cinema management division into a new subsidiary called Lotte CultureWorks.[57]

Megabox is South Korea's third major exhibitor and notable because its origins lie in the only incursion into the South Korean cinema market by a US exhibitor—Loews Cineplex. In 2000, the US company formed a joint venture with Mediaplex, a subsidiary of confectionary company and conglomerate the Tongyang Group, with the brand name Megabox. Its first cinema—the Megabox COEX—opened in May 2000, in the COEX mall, adjacent to Seoul's World Trade Centre, and with 16 screens and 4336 seats it was then South Korea's largest cinema. Loews Cineplex's influence in the venture was particularly evident in the focus on concessions, and popcorn in particular, as an important driver of revenues, something that was traditionally less important in South Korean cinemas.[58] The Megabox COEX was followed by a series of smaller multiplexes in several other cities, along with creating a second tier of cinemas branded as Megaline, opened in smaller towns and cities. By the end of 2004 there, the company operated seven Megabox multiplexes and six Megaline cinemas. As one of the busiest cinemas in the country, the Megabox COEX contributed to a reported 15 million admissions to the circuit's cinemas in 2002.[59]

Loews Cineplex's investment in Megabox lasted only five years, however, when in December 2005 it sold 45 per cent of its share to the British investment bank Standard Chartered, and 5 per cent to Mediaplex. The sale was linked to the imminent merger between Loews Cineplex and AMC in the USA. With Village Roadshow's exit from its joint venture in CGV in 2002, this meant that there was no presence in the South Korean market by any overseas exhibitor; though in 2007 Mediaplex sold the circuit to a consortium called Korea Multiplex Investment Corporation, which was supported by Australian bank Macquarie, though Mediaplex continued to manage the cinemas.[60] Megabox has been traded several times since and in 2017 was owned in part by Shinhan BNP Paribas. Nonetheless, despite all of the changes in ownership, the circuit has continued to grow and in 2017, it had 95 cinemas with 650 screens.[61]

The rapidity with which South Korea adopted the multiplex cinema has no obvious parallel, with the penetration of multiplexes (defined by the Korean Film Council for statistical purposes as a cinema with seven or more screens) accounting for more than three quarters of screens. Given that in 1998 all of South Korea's cinemas were single-screen, this is a remarkable transformation in just two decades.[62] Moreover, unlike Japan, for instance, more than half of all visits to the cinema are to see domestically produced films.[63] Perhaps the most significant indication of the development of South Korea's exhibition industry is the fact that in the late 1990s cinema attendance per capita stood at one visit per year, but by 2017 South Koreans went to the cinema more than any other country in the world—4.27 visits per year.[64]

## CHINA

Considering China is a highly centralised and bureaucratised state, ascertaining the number of cinemas is difficult. As *Screen Digest* observed in 1992, "although figures are well reported, there are several versions to choose from."[65] In part, this has been due to the high proportion of mobile cinemas, touring rural areas, which would not constitute 'cinemas' in the context of this study, whilst over-reporting by official sources has been endemic, historically. So, for instance, in 2000, official figures from the now-defunct State Statistical Bureau suggested that there were 65,000 screens in China; however, *Screen Digest* calculated that less than a third of these were screens in city cinemas and very small proportion were showing imported films.[66] Most analysts subsequently consider China's cinema infrastructure in terms of screens in urban areas, that is, those located in modern cinemas, in China's major cities. On this basis, in 2004, there were 1188 cinemas with 2396 screens, which had tripled by 2012 to 3680 sites with 13,118 screens, suggesting extensive multiscreening.[67] By 2017, China had undergone an exponential growth with 9169 cinemas and over 50,000 screens,[68] whilst in 2018 a further 9303 screens were opened.[69] In terms of the growth in multiplex cinemas, Peichi Chung and Lianyuan Yi calculated that there were 1188 in 2004, which had risen to 5598 ten years later in 2014.[70] Allowing for a whole series of abuses and issues in the reporting of box-office, it is clear that China's market has grown, 33 per cent a year on average since 2002.[71]

The first multiplexes began to emerge in China in the late 1990s, with the first being a five-screen, 1218-seat site in Chongqing, opened by SMILE (SMI Leisure and Entertainment Ltd), a consortium of Shanghai-based Cai Wei, South Malaysian Industries (SMI), and British exhibitor United Cinemas International (UCI). Cai Wei, which provided the real estate, held approximately 60–70 per cent of the deal, with UCI providing "technical services," contributing 5 per cent.[72] The consortium, which had some $60 million to invest, was formed three years earlier, with UCI keen to enter what they felt was at best a moribund market but with huge growth potential.[73] At that point, UCI were the first Western exhibition company to commit to enter the Chinese market, albeit in such a limited capacity.

UCI's investment was small compared to the second major Western entrant into the market: Warner Bros. International Theaters (WBIT); which in 2002 took a 49 per cent stake in a joint venture with a local circuit called Shanghai Paradise, to open the nine-screen Paradise Grand Gateway multiplex in Shanghai; making WBIT the first foreign exhibitor in China.[74] Shanghai Paradise was the city's leading exhibitor, operating 140 screens, representing 90 per cent of the total commercial screens in Shanghai.[75] Shanghai Paradise's association with Warner Bros. had begun in 1992, when they were able to help facilitate a distribution agreement with the China Film Group in 1994: the first to be signed by a major Hollywood studio in China.[76]

This was the first of several joint venture deals signed by WBIT, spurred on in no small part by the Chinese government's efforts to open up the film market to foreign investment, largely by changing the regulations to allow foreign investors to hold up to 75 per cent of the capital in exhibition chains in six major cities—Beijing, Shanghai, Guangzhou, Chengdu, X'ian, Wuhan, and Nanjing.[77] The growth in the numbers of screens in modern cinemas from the late 1990s onwards in China was largely accounted for by state-owned, vertically integrated film bodies such as the China Film Group and Shanghai Film Group Corporation, which invested in local exhibitors or entered into joint ventures with overseas companies. WBIT's most significant agreement was signed with Shanghai Film Group Corporation in 2004, to invest in ten multiplexes, with the first sites opened in Nanjing, Tianjin, and Wuhan. However, in 2005, WBIT sold out its stakes in its various joint ventures to its partners, in the face of the reversal by the Chinese government of its regulatory change, which now stated that Chinese investors must once again hold 51 per cent in any joint venture with a foreign investor.

Whilst the entry of companies like UCI and WBIT suggested the partial liberalisation of the Chinese market, their presence was considerably less extensive than that of other regional exhibitors, especially those based in Hong Kong and South Korea. Lark International Entertainment, whose Chinese exhibition arm Studio City Cinemas (SCC) was one of the first multiplex operators in China, opened three five- and six-screen multiplexes in Wuhan, Chongqing, and Shanghai between 1997 and 1999. Lark had opened a multiplex cinema in Hong Kong, in a joint venture with the US United Artists Theatre Circuit (UATC) in 1985, before buying out its partner in 1997, whereupon it operated eight cinemas with 30 screens.[78]

When former British colony Hong Kong was returned to China in 1997, becoming a Special Administrative Region, Lark was able to access the Chinese mainland. Similarly, Golden Harvest entered the Chinese multiplex market in 2004, when it opened a seven-screen site in Shenzhen, on the Chinese mainland across the border from Hong Kong. As a Hong Kong-based company, Golden Harvest, like Lark, was able to take advantage of the Closer Economic Partnership Arrangement (CEPA), a free trade agreement concluded between mainland China and Hong Kong in June 2003. This allowed them to control majority stakes in joint ventures unlike foreign companies. Golden Harvest expanded in China and in 2007, Chinese company Orange Sky Entertainment acquired a controlling share in the company, renaming it Orange Sky Golden Harvest (OSGH). The company had opened 42 sites with 306 screens by 2012[79] and by 2016 it had 76 sites and 531 screens, whereupon it sold its sites to China's second largest domestic exhibitor Guangdong Dadi.[80] As we have seen, neighbouring South Korea has a presence in the Chinese exhibition market, with Lotte CultureWorks, Megabox, and CGV all operating multiplexes, though with 100 sites, CGV is considerably bigger than its rivals.

In any event, the operations of overseas circuits are small indeed when compared to the major domestic circuits that have emerged, though, as Dodona pointed out, the precise details of ownership and, indeed, screen counts are not always altogether clear.[81] Nonetheless, China's largest circuit in 2017, by some margin, was Wanda Line Cinema Corporation, with 525 cinemas and 4680 screens.[82] Owned by Dalian Wanda, Wanda Line Cinema is also an example of investment flowing out of China rather than in, since it now owns AMC and Carmike in the USA, the ODEON Group in Britain and Europe, the Nordic Cinema Group in Scandinavia, and Hoyts in Australia. It is the largest exhibition company in the world with 1351 cinemas and over 14,000 screens.[83] Dalian Wanda opened China's largest cinema in 2018—a 30-screen, 5000-seat megaplex, located in a shopping mall in its Quingdao Movie Metropolis; the world's biggest film studio complex. China's next largest exhibitor is Guangdong Dadi with 464 cinemas and 2745 screens, followed by China Stellar with 365 cinemas and 2290 screens.[84]

Examining the numbers of sites and screens suggests that the average multiplex in China is considerably smaller than those in the USA, Britain, and other territories, with five to six screens being the norm. The cycle of openings and closures is high too, which is a function of the extremely competitive nature of the market, in which cities can have a large number of multiplexes in close proximity. In 2014, von Sychowski highlighted the case of Hangzhou, a provincial city in in the Hangzhou Metropolitan Area, as one in which some 49 cinemas had opened, with ten more planned

in 2015.[85] The opening of large numbers of multiplexes in provinces outside of the major urban centres (Beijing, Shanghai, and Guangdong, for
instance), like those of Hangzhou, has been driven by a demand for new
markets, though in these provinces the ticket price is lower since earnings
are lower.

Notwithstanding the differential pricing, in China, going to the multiplex is largely the domain of the urban, middle classes, a result of the
emergence in China of a focus on domestic consumption, commercial real
estate, and the growth of the city. As Yi Lu observed:

> China's entertainment and leisure industries have had to converge with the
> international trend of satisfying the needs of an emerging middle class long-
> ing to participate in metropolitan modernity and consumerism. In particu-
> lar, the state's deregulation policies on cinema can be understood as a
> continuation of its leisure consumption policy of the mid-1990s and as a
> concrete action of converting cultural capital into economic capital.[86]

This is reflected in the growth of premium cinemagoing experiences,
such as IMAX, which has made a substantial investment in the Chinese
market, with 543 installations and more than 300 contracted to open in
2018.[87] China's burgeoning cinema market was also one of the first and
fastest to instal digital projection, led by the state-owned China Film
Group, which began the scheme in 2004 with a plan to instal digital
90–100 screens. In order to promote developments, the government
exempted the import of digital films from the official film quota, which
meant that with higher prices charged the box-office of films like *Star
Wars: Episode II—Attack of the Clones* and other digital titles took a larger
proportion of the box-office.[88] The installation of the DCI-compliant projectors was rapid too, with 93 screens in 2004, increasing to 652 in 2007,
including 82 equipped for 3D and 12,225 in 2012, with 7200 3D.[89] By
2017, digital screens had become universal.

Since joining the World Trade Organisation on 11 December 2001,
and losing a ruling in 2009 on the state's monopolisation of the import
and distribution of foreign films, China has steadily become one of the
largest film markets in the world. All the industry forecasts are that the
exhibition sector will continue to expand, driving the Chinese market's
inexorable growth, so that within two or three years it will assume the
position as the world's number one in terms of box-office. Based upon
China's rate of cinema construction and openings, the number of screens,

already the highest in the world, will continue to grow. However, even in assuming its position as the biggest market, it is worth reflecting on the fact that the figure for cinema attendance per capita of 1.16 is extremely low in comparison to most other countries, making the potential for growth considerable.

## NOTES

1. Philip Singleton, Executive VP of AMC, quoted in Alan Waldman, "The Road to Asia," *The Hollywood Reporter Salute to Showester of the Year Stan Durwood* (4 March, 1996), S-10.

2. *MPAA, 2018 Theme Report* (Motion Picture Association of America Inc., 2019), 9. https://www.mpaa.org/wp-content/uploads/2019/03/MPAA-THEME-Report-2018.pdf (accessed 10 April, 2019).

3. Dodona, *Cinemagoing Asia Pacific 2004* (Leicester: Dodona Research, 2004), 4.

4. MPAA, *2018 Theme Report*, 10.

5. Dodona, *Cinemagoing Asia Pacific 2004*, 9.

6. Korean Film Council, *Status & Insight: Korean Film Industry 2017* (Korean Film Council, Film Policy Research Institute, 2017), 11. www.koreanfilm.or.kr/eng/publications/download.jsp?fileNm=Theme413.pdf (accessed 29 January, 2019).

7. Motion Picture Producers Association of Japan, Inc., *Statistics of Film Industry in Japan*, http://www.eiren.org/statistics_e/index.html (accessed 10 January, 2019).

8. Dodona, *Cinema Industry Research China, Hong Kong, Taiwan 2018* (Leicester: Dodona Research, 2018), 1.

9. Ibid., 1.

10. "WB: Joint Venture, Per Capita," *BoxOffice* 127, no. 7 (1 July, 1991), 24.

11. Garth Alexander, "WB, Grocer to Build Japan Multiplexes," *Variety* 343, no. 5 (13 May, 1991), 6.

12. Michelle Magee, "Multiplexes Moving In," *Variety* 364, no. 7 (16 September, 1996), 60.

13. Ibid., 60.

14. Motion Picture Producers Association of Japan, Inc., *Statistics of Film Industry in Japan*.

15. Ibid.

16. Dodona, *Cinemagoing Asia Pacific 2004*, 127.

17. Motion Picture Producers Association of Japan, Inc., *Statistics of Film Industry in Japan*.

18. Dodona, *Cinemagoing Asia Pacific 2004*, 130.

19. UNIJAPAN, *The Guide to Japanese Film Industry & Co-production 2009* (Tokyo: UNIJAPAN, 2009), 14.

20. Dodona, *Cinema Industry Research Japan 2018* (Leicester: Dodona Research, 2018), 3.

21. Millard Ochs, "Plexing muscle," *Screen International* 1150 (20 March, 1998), 26.

22. Notwithstanding occasional exceptional years, such as 1997 when *Princess Mononoke* dominated the box-office, when one considers the trends, domestic films account for approximately 35 per cent of the film market in Japan.

23. Dodona, *Cinemagoing Asia Pacific 2004*, 138.

24. Quoted in Waldman, "The Road to Asia," S-10.

25. Jon Herskovitz, "Distribs Maintain Hold on Hits Via Archaic Booking Systems," *Variety* 371, no. 2 (18 May, 1998), 58.

26. Scott Rosenberg, "AMC Comes to Asia: U.S. Circuit Finds Success in far East," *Film Journal International* 102, no. 3 (1 March, 1999), 48.

27. AMC Entertainment Inc., *1997 Annual Report* (Leawood, KS: AMC Entertainment Inc., 1997), 16.

28. Rosenberg, "AMC Comes to Asia," 48.

29. Alan McNair, UCI's chief financial officer, quoted in Beth Porter, "UCI Looks East," *Film Journal International* 100, no. 1 (1 February, 1997), 30.

30. Dodona, *Cinemagoing Asia 2007* (Leicester: Dodona Research, 2007), 47.

31. Andreas Fuchs, "Tea Houses and Multiplexes," *Film Journal International* 100, no. 1 (1 February, 1997), 46.

32. Dodona, *Cinema Industry Research Japan 2018*, 4.

33. Dodona, *Cinemagoing Asia Pacific 2004*, 140–1.

34. Mamiko Kawamoto, "T Joy Opens Second E-Cinema Complex in Japan," *Variety* (31 May, 2001). https://www.screendaily.com/t-joy-opens-second-e-cinema-complex-in-japan/405849.article (accessed 30 January, 2019).

35. Dodona, *Cinemagoing Asia 2008* (Leicester: Dodona Research, 2008), 25.

36. Dodona, *Cinemagoing Asia 2012* (Leicester: Dodona Research, 2012), 19.

37. Ibid., 19.

38. Dodona, *Cinema Industry Research Japan 2018*, 4.

39. Motion Picture Producers Association of Japan, Inc., *Statistics of Film Industry in Japan*.

40. Korean Film Council, *Status & Insight: Korean Film Industry 2017*, 55.

41. Ibid., 54.

42. Ibid., 27
43. "Cinemas and Their Audiences: Just holding on," *Screen Digest* (September, 1992), 206.
44. Ibid., 203.
45. Sangho Kim, "Cinema Demand in Korea," *Journal of Media Economics* 22, no. 1 (March, 2009): 36–56.
46. Yeongkwan Song, *Audiovisual Services in Korea: Market Development and Policies*. ADBI Working Paper 354. (Tokyo: Asian Development Bank Institute, 2012). http://www.adbi.org/working-paper/2012/04/16/5048.audiovisual.services.korea/ (accessed 29 January, 2019).
47. Dodona, *Cinemagoing Asia Pacific 2000* (Leicester: Dodona Research, 2000), 93.
48. Sohee Kim, "Foreign Correspondence: Korea Scene," *Film Journal International* 101, no. 4 (1 April, 1998), 128.
49. Cheil Jedang was an early investor in new US studio Dreamworks SKG, buying 11 per cent of the stock for $300 million in return for the rights to distribute the company's films in Asia outside Japan for ten years.
50. Dodona, *Cinemagoing Asia Pacific 2004*, 230.
51. Christopher Alford, "CJ Village at home in Korean exhib'n," *Variety* 380, no. 13 (13 November, 2000), 84.
52. Lee Hyo-won, "South Korea's CJ CGV to Invest $200M in Vietnamese Theaters by 2020," *The Hollywood Reporter* (9 April, 2017), https://www.hollywoodreporter.com/news/south-koreas-cj-cgv-invest-200m-vietnamese-theaters-by-2020-1034929 (accessed 30 January, 2019).
53. Dodona, *Cinema Industry Research South Korea 2018* (Leicester: Dodona Research, 2018), 5–6. CGV has a 40.25 per cent interest in Turkish circuit Mars Entertainment and a 38.1 per cent interest in its Indonesian CGV circuit.
54. Korean Film Council, *Status & Insight: Korean Film Industry 2017*, 54.
55. Ibid., 6.
56. Kim Anh, "Foreign Players dominate blossoming cinema scene," *Vietnamese Investment Review* (28 September, 2018). https://www.vir.com.vn/foreign-players-dominate-blossoming-cinema-scene-62707.html (accessed 30 January, 2019).
57. Patrick Frater, "Korea's Lotte Restructures Film Ahead of Overseas Push, Streaming Launch," *Variety* (10 April, 2018). https://variety.com/2018/biz/asia/korea-lotte-restructures-film-creativeworks-1202748912/ (accessed 29 January, 2019).
58. Christopher Alford, "Screening for quality," *Variety* 378, no. 10 (24 April, 2000), 56.
59. Dodona, *Cinemagoing Asia Pacific 2004*, 232.
60. Dodona, *Cinemagoing Asia 2008*, 49.

61. Korean Film Council, *Status & Insight: Korean Film Industry 2017*, 56.
62. Dodona, *Cinema Industry Research South Korea 2018*, 4.
63. Korean Film Council, *Status & Insight: Korean Film Industry 2017*, 11.
64. Dodona, *Cinema Industry Research South Korea 2018*, 1.
65. "Cinemas And Their Audiences: Just holding on," 206.
66. "Worldwide Cinema: Key markets strong as global admissions dip," *Screen Digest* (October, 2001), 314.
67. Dodona, *Cinema Industry Research China, Hong Kong, Taiwan 2013* (Leicester: Dodona Research, 2013), 5.
68. Dodona, *Cinema Industry Research China, Hong Kong, Taiwan 2018*, 3.
69. Rebecca Davis, "China Box Office Growth Slows to 9% in 2018. Hits $8.9 Billion," *Variety* (2 January, 2019). https://variety.com/2019/film/news/china-box-office-2018-annual-1203097545/ (accessed 4 February, 2019).
70. See Peichi Chung and Lianyuan Yi, "The regionalization of co-production in the film industries of Hong Kong SAR and mainland China," in *Handbook of Cultural and Creative Industries in China*, ed. Michael Keane (Cheltenham: Edward Elgar, 2016), 219.
71. Dodona, *Cinema Industry Research China, Hong Kong, Taiwan 2018*, 1.
72. Scott Rosenberg, "Reason to Smile," *Film Journal International* 100, no. 4 (1 May, 1997), 52.
73. Mark Schilling, "UCI pits Smile into Chinese exhibition," *Screen International* 952 (8 April, 1994), 4.
74. "Global Cinema Exhibition: Record Revenues But Dollar Values Mask Local Currency Falls," *Screen Digest* (September, 2004), 276.
75. Warner Bros., *Warner Bros. International Theatres to Open Multiplex Cinema in Shanghai With Local Partners*. Press Release, 4 March 2002. https://www.warnerbros.com/studio/news/warner-bros-international-theatres-open-multiplex-cinema-shanghai-local-partners (accessed 23 December, 2018).
76. Ibid.
77. "Focus On Emerging Markets: China," *Screen Digest* (September, 2004), 276.
78. Melissa Morrison, "Having a Lark," *BoxOffice* 134, no. 12 (1 December, 1998), 24.
79. Dodona, *Cinema Industry Research China, Hong Kong, Taiwan 2013*, 7.
80. Patrick Frater, "Dadi Pays $575 Million for Orange Sky Golden Harvest's China Cinemas," *Variety* (26 January, 2017). https://variety.com/2017/biz/asia/dadi-pays-575-million-for-orange-sky-golden-harvests-china-theaters-1201970458/ (accessed 4 February, 2019).
81. Dodona, *Cinema Industry Research China, Hong Kong, Taiwan 2018*.
82. Ibid., 4.

83. Wanda group, *Overseas Film companies*, http://www.wanda-group.com/movies/ (accessed 26 January, 2019).
84. Dodona, *Cinema Industry Research China, Hong Kong, Taiwan 2018*, 4.
85. Patrick von Sychowski, *Does Hangzhou Point to Bubble in China's Cinema Market?* Celluloid Junkie (19 November, 2014) https://celluloidjunkie.com/2014/11/19/hangzhou-point-bubble-chinas-cinema-market/ (accessed 5 February, 2019).
86. Yi Lu, "The Malling of the Movies: Film Exhibition Reforms, Multiplexes, and Film Consumption in the New Millennium in Urban China," *Journal of Chinese Cinemas*, 10, no. 1 (2016): 205–227, 206.
87. Patrick Frater, "IMAX Signs 30-Theater Deal With China's Jin Yi media," *Variety* (3 April, 2018). https://variety.com/2018/film/asia/imax-30-theater-deal-with-chinas-jinyi-media-1202742224/ (accessed 7 February, 2019).
88. Dodona, *Cinemagoing Asia Pacific 2004*, 67.
89. Dodona, *Cinema Industry Research China, Hong Kong, Taiwan 2013*, 5.

# CHAPTER 12

# Conclusion: The Multiplex as a Global Form

This book has covered eight of the largest film markets in the world in 2017 by gross box-office, including six of the largest film markets by admissions.[1] The growth in cinema screens since the first multiplexes began to appear in the 1980s has been considerable, with *Screen Digest* estimating that in 1980 there were just under 86,000 worldwide.[2] In 2017, according to the Motion Picture Association of America (MPAA), there were 171,755 screens globally,[3] of which 55 per cent were located in China and the USA.[4] If one considers all of the countries covered in this study, they account for approximately 65 per cent of the global screen count, though given the maturity of many of the markets they account for a considerably higher proportion of multiplex screens.

In the USA, just under 36,000 screens are in multiplexes, which is 89 per cent of the total,[5] whilst the total for China is less clear, though it is reasonable to suppose that given the fact that only new urban screens are counted, then the total screen numbers are located in multiplexes; albeit smaller ones than is the norm in other territories. In the countries with high rates of screens and attendance, admission rates per capita are low and the diffusion of the multiplex is much less extensive. India is a case in point, with 9530 screens for a population in excess of 1.3 billion, meaning that Indians go to the cinema just over once every two years per capita.[6] Multiplexes, which tend to average three to five screens, are located predominantly in larger urban centres and generally the reserve of India's middle class. In 2017, there were 2750 multiplex screens in the context of

© The Author(s) 2019      253
S. Hanson, *Screening the World*,
https://doi.org/10.1007/978-3-030-18995-2_12

6750 single-screen cinemas.[7] The potential for growth is substantial but fraught with obstacles such as affordability and high land rents for exhibitors; whilst the sector is dominated by four exhibitors, the biggest being PVR (a joint venture between Priya and Australian exhibitor Village Roadshow) and Cinepolis, Mexico's largest exhibiter. Nevertheless, within established markets—USA, Britain, and Western Europe—the scope for expansion is considerably less than that taking place in some countries in South-East Asia (Indonesia and Malaysia), Africa (Nigeria and Egypt), and Latin America (Brazil and Mexico). In Latin America, for instance, screen growth in key markets has risen by nearly 50 per cent in the last decade.[8]

The increase in the number of multiplexes has been driven largely by two stages of consolidation: regional and global. In many markets, such as South Korea, Australia, and Britain, exhibitors have expanded regionally through a process of acquisition and merger, as a way of accessing markets and growing. When they do, they take with them an increasingly standardised model of a multiplex, in terms of design, operation, and film programming. Increasingly, the global multiplex industry is also consolidating, with the emergence of first large domestic, then regional, and now transnational circuits, which are increasingly expanding also via a process of acquisition and merger. Primary amongst these, as we have seen, is the Chinese conglomerate Dalian Wanda, which, through its subsidiary Wanda Cinema Line, now owns exhibitors in China, USA, Britain and Europe, Scandinavia, and Australia, along with Hollywood film company Legendary Pictures. Nine other companies have a substantial presence in the global exhibition market and whose circuits are characterised by multiplexes: Cineworld (Britain), Cinemark (USA), Cinépolis (Mexico), CGV (South Korea), Cinemex (Mexico), Vue (Britain), Cineplex Entertainment (Canada), Pathé (France), and Kinepolis (Belgium). One of the ambitions of these large, trans-national exhibition companies is to be able to deal directly with the major studios. As Hancock observed a "world in which three exhibitors control half the world's screens (which could feasibly happen) would be a very different world for distributors, manufacturers, and fellow exhibitors."[9]

The pattern of mergers and acquisitions has been largely driven by a relentless quest for economies of scale and the associated cost savings as a route to greater profitability; though in many cases the acquisitions have relied on substantial amounts of leveraging and, therefore, debt. As circuits expand, the barriers to entry for rivals in their domestic markets become greater, but this also means that they have to adopt a defensive posture in order to protect those markets; whilst at the same time looking to expand into new ones. Writing in 1999, Russell Scott, General Manager of Australia's Greater Union, commented that:

> There are a number of regions in the world that are underscreened. Australian cinemas are recognized to have high standards and are considered experts in multiplex development. This is a competitive advantage for the Australian chains and is a skill and knowledge that is exportable to other countries. With the limitations in future growth offered by the Australian cinema market, each of the majors has researched overseas opportunities and has adopted strategies for globalization. These strategies involve taking their cinema expertise (their core business) and establishing it in other countries.[10]

It is inconceivable now that anyone opening a cinema would not open a multiplex, or at the very least a multiscreen cinema. For exhibitors, the multiplex has come to symbolise a particular form of industrial organisation in which the imperative is to maximise operational efficiencies; since with lots of screens but only one or two projection rooms and the concession sales increasingly utilising self-service, capital and labour costs can be greatly reduced. Moreover, with multiple screens of differing sizes, exhibitors can manage the run of any film in order to maximise returns. In terms of product, multiplex operators, like all exhibitors, rely on a combination of big-budget, largely US-produced, films and, depending on the territory, a number of domestic films. The dependence on the major Hollywood studios and their affiliated distribution networks for films that are able to generate large audiences means that negotiating bilateral agreements with what is effectively an oligopoly is key. When one considers the top ten films at the global box-office in 2018, all were US produced or US co-productions, with four films alone taking nearly $7 billion.[11] Few countries have a domestic industry that takes 50 per cent or above of the film market, especially where they have exhibitors which are not part of vertically integrated film companies. For all circuits in all places the multiplex has a symbiotic relationship to the blockbuster, foreign or domestic. Distributors have come to utilise the multiplex to open films widely, boosting opening weekends, and maximise marketing potential, with simultaneous global releases now the norm. As Cucco argued, multiplexes:

> allow perfect management of the movie during its life-cycle. When a film begins to lose viewers, it can be progressively moved to smaller theatres,

keeping the movie in the portfolio and rationalizing the use of the spaces in relation to demand. It is difficult to establish whether this strategy is a cause or a consequence of the choice of concentrating takings on the first weekend; however, multiplex cinemas appeared before the development of this distribution strategy. Consequently, it is possible that distributors have found in these new cinemas a tool suited to their purpose.[12]

Thus, to watch *The Fast and the Furious* in a multiplex in Tokyo, Seoul, Madrid, Abu Dhabi, Ho Chi Minh City, Mexico City, or Abuja is increasingly to be part of a global cinema 'experience.' The increasing homogeneity on the screen is matched too by the multiplex environment both technologically and aesthetically. As multiplexes in many countries have sought to distinguish their film-watching offer from that of the home, this means it is increasingly marketed as a luxury event, with widespread upgrading of auditoria and the installation of ever more sophisticated forms of film-viewing technologies. These technologies, which are increasingly marketed by globalised companies, include digital 3D, 4DX, laser projection, Dolby Atmos sound, and IMAX, in addition to the newest innovation of flat panel cinema screen composed of LED panels, which replaces the traditional screen and projector. The developer of this technology, Samsung, has struck a series of deals in Asia-Pacific with exhibitors like Wanda and Lotte. With luxury cinemas come luxury prices, with many multiplexes, particularly in emergent countries, the domain of the more affluent.

Multiplexes have also come to be closely identified with the development of leisure and retailing, and particular kinds of retail and leisure spaces: the mall. These are familiar spaces to ever-growing numbers of people around the world. In Manila, in the Philippines, Schilders highlighted that:

> [i]t is easy to go to the cinema, because every mall has one. Manila counts over thirty cinemas, most of which are multiplexes, if not megaplexes. Although the majority of the population in Manila lives from 5000 Philippine pesos (about 107 USD) a month, cinema-going is popular. An average meal costs 50 or 60 Php. A ticket for the cinema is 130 Php. A beer in a "nice place" costs 100 Php as well. But movies have the magic factor, letting you forget about your daily life for the time being.[13]

The multiplex is, thus, an overwhelmingly urban or perhaps a (sub) urban form, closely linked with modernity. As Athique and Hill observe:

"[i]n that sense, the multiplex is held as something new in that it offers a range of choices oriented around our contemporary understanding of what it is to be modern and to be pleasured."[14]

If this book had been written in the 1980s, it would have cited the Video Cassette Recorder (VCR) as the perceived main threat to cinema, and in the 1990s it would have been the DVD. In 2018, with the demise of the videocassette and the decline of the DVD and Blu-Ray, it is streaming services like Netflix, which now constitute the challenge to cinema in terms of film viewing. Netflix has 139 million subscribers globally,[15] passing the symbolic 50 million mark in the USA, which seems to suggest, when one considers the 75 million subscribers for Amazon Prime Video,[16] that streaming is an existential threat to the multiplex. However, the increased box-office in 2018, largely on the back of a small number of films including *The Black Panther*, *Avengers: Infinity War*, and *The Incredibles 2*, seemed to suggest that this view was premature. Nonetheless, there exist a set of contrasting views: that the relationship between streaming services like Netflix and the cinema is a mutually beneficial one or that the increase is a blip and that the fundamental shift is away from the cinema towards home-viewing. According to the US National Association of Theater Owners (NATO), which conducted a study in 2017 amongst 2002 cinemagoers, "[t]hose who attended movies in theatres more frequently also tended to consume streaming content more frequently."[17] The opposing view on the threat of Netflix and others was articulated by one anonymous film executive cited by Zeitchik: "I don't think anyone who looks at the challenges movies face in a crowded landscape could argue Netflix is helping theatrical box office," adding, "for most screen content the experience at home isn't that different from the one in a theater."[18]

Netflix's co-founder and CEO Reed Hastings had once professed that Netflix's main competitor was not another media entity but rather the need to sleep,[19] though he was also keenly aware of the kudos afforded to their original film content by the nomination for awards. With acquisition of film content and latterly film production, a key element of the company's strategy, it has looked to release some of its titles in cinemas a few weeks before they are available to subscribers, as in the case of Alfonso Cuarón's *Roma*. However, many of the large circuits refused to screen it, claiming that its imminent release on Netflix's streaming service undermined the traditional windowing structure, in which studios wait for at least 90 days or so, before making their film available for viewing in the home. "We're not pro-theater, we're not anti-theater," Netflix's Ted Sarandos, argued, "we're pro-consumer," though NATO countered that they (Netflix) "want all the benefits of a theatrical release, and they don't want any of the risks."[20] For many analysts, though, the threat is actually to Netflix, since they have registered huge cash losses of $13 billion since 2011, with subscriber growth not generating enough revenue to make up for the huge costs in production of content.[21]

With several of the major Hollywood studios planning their own streaming services and Disney+ due to begin operation in 2019, the volatility in the market might allow the space for the cinema after all. Moreover, for many young people wedded to their smartphones and tablets, the cinema continues to be important: "[d]on't count cinema out though, not by a long shot," argued Arnold, "movie theaters bring something that streaming cannot: an immersive experience. That's something that Millennials crave."[22]

To argue for the importance of the multiplex in the reinvigoration of cinema attendance in all of the territories discussed is not, however, to deny that they constitute a powerful economic and cultural force in cinema exhibition that might not be seen as wholly positive. Their presence in the cinema landscape has to be constantly qualified, whilst the story of the multiplex is one of concentration of ownership in all of the regions in which they have emerged. They have trumpeted "choice," yet the evidence in many instances is homogeneity in terms of films shown with a concomitant reliance on films from the major Hollywood studios and distributors. Nonetheless, as we approach the end of the second decade of the twenty-first century, we might see the enduring value in the projection of feature films to a paying audience in a communal space.

## Notes

1. European Audiovisual Observatory, *Focus 2018: World Film Market Trends* (Marché Du Film: Festival De Cannes, 2018), 11.
2. "Cinemas and their Audiences: Just holding on," *Screen Digest* (September, 1992), 203.
3. MPAA, *2017 Theme Report* (Motion Picture Association of America Inc., 2018), 9. https://www.mpaa.org/research-docs/2017-theatrical-home-entertainment-market-environment-theme-report/ (accessed 7 December, 2018).
4. European Audiovisual Observatory, *Focus 2018: World Film Market Trends*, 12.

5. Dodona, *Cinema Industry Research: USA & Canada 2018* (Leicester: Dodona Research, 2018), 6.
6. Dodona, *Cinema Industry Research India 2018* (Leicester: Dodona Research, 2018), 6.
7. Ibid., 3.
8. David Hancock, "Latin America Cinema Markets On The Way Up," *Cinema Technology* 30 no. 4 (December, 2017), 24.
9. David Hancock, "Transforming the Movie Experience," *BoxOffice Pro: The Giants of Exhibition 2017* (February, 2017), 26. https://pro.boxoffice.com/wp-content/uploads/2017/02/Giants2017.pdf    (accessed    11 February, 2019).
10. Russell Scott, "Exporting Expertise," *Film Journal International* 102, no. 7 (1 July, 1999), 16.
11. Box Office Mojo, *2018 Worldwide Grosses*. https://www.boxofficemojo.com/yearly/chart/?view2=worldwide&yr=2018&p=.htm    (accessed    10 February, 2019).
12. Marco Cucco, "The Promise is Great: The Blockbuster and the Hollywood Economy," *Media, Culture & Society* 312 (2009): 215–30, 227–8.
13. Emmy Schilders, "The Experience of Cinema Going in Manila [Philippines]," *Senses of Cinema-Going: Brief Reports on Going to the Movies Around the World*, eds. Arthur Knight, Clara Pafort-Overduin, and Deb Verhoeven (March, 2011). http://sensesofcinema.com/2011/feature-articles/senses-of-cinema-going-brief-reports-on-going-to-the-movies-around-the-world/ (accessed 12 February, 2019).
14. Adrian Athique and Douglas Hill, *The Multiplex in India: A Cultural Economy of Urban Leisure* (London: Routledge, 2010).
15. Natalie Jarvey, "Netflix Grows Subscriber Base to 139 Million Worldwide," *The Hollywood Reporter* (17 January, 2019). https://www.hollywoodreporter.com/news/netflix-grows-subscriber-base-139-million-worldwide-1176934 (accessed 12 February, 2019).
16. Todd Spangler, "Amazon's Prime Video Channels Biz to Generate $1.7 Billion in 2018," *Variety* (7 December, 2018). https://variety.com/2018/digital/news/amazon-prime-video-channels-tv-revenue-estimates-1203083998/ (accessed 13 February, 2019).
17. Phil Contrino, "Ernst & Young Survey Finds Frequent Moviegoers are also Frequent Streamers," *NATO News Reel Blog* (24 April, 2018). http://www.natoonline.org/blog/reel-blog/ernst-young-survey-finds-frequent-moviegoers-are-also-frequent-streamers/ (accessed 29 December, 2018). Ernst and Young report, *The Relationship between Movie Theatre Audiences and Streaming Behaviour: Survey Insights*, April, 2018. www.natoonline.org/wp-content/uploads/2018/04/The-relationship-between-movie-theatre-attendance-and-streaming-behavior-4-26-18.pdf    (accessed    23 January, 2019).

18. Steven Zeitchik, "Is Netflix Killing the Movie Theater? Not so Fast," *The Washington Post* (24 December, 2018). https://www.washingtonpost.com/business/is-netflix-killing-the-movie-theater-not-so-fast/2018/12/24/7a1 6dbf8-037a-11e9-8186-4ec26a485713_story.html?noredirect=on&utm_term=.273d927a5d93 (accessed 28 December, 2018).

19. Charles Bramesco, "Is Netflix About to Change How It Releases Original Movies?," *The Guardian*, 31 August, 2018.

20. Ryan Faughnder and Josh Rottenberg, "What Netflix's Release of 'Roma' Says about Its Movie Business Strategy," *Los Angeles Times* (3 December, 2018). https://www.latimes.com/business/hollywood/la-fi-ct-netflix-roma-20181203-story.html (accessed 23 December, 2018).

21. David Trainer, "Reality is Closing In On Netflix," *Forbes* (5 February, 2019). https://www.forbes.com/sites/greatspeculations/2019/02/05/reality-is-closing-in-on-netflix/#1c25cfc5ed0b (accessed 13 February, 2019).

22. Andrew Arnold, "Convenience Vs. Experience: Millennials Love Streaming But Aren't Ready to Dump Cinema Just Yet," *Forbes* (26 October, 2017). https://www.forbes.com/sites/andrewarnold/2017/10/26/millennials-love-streaming-but-arent-ready-to-dump-cinema-just-yet/#6d6532 f46311 (accessed 13 February, 2019).

# Select Bibliography

"$120,730,000 Invested in New Theatres." *BoxOffice* 90, no. 13 (16 January, 1967).

"$131,122,708 in 450 New Theatres, 221 Located In Shopping Centers." *BoxOffice* 86, no. 13 (18 January, 1965).

"$147,836,000 in 454 Theatres, 238 Located in Shopping Centers." *BoxOffice* 88, no. 13 (17 January, 1966).

"$54,725,400 Invested in New Theatres in 1961." *BoxOffice* 81, no. 3 (7 May, 1962).

"$90,706,500 For 242 New Theatres During '62." *BoxOffice* 82, no. 14 (4 February, 1963).

"$97,411,500 Invested in 320 New Theatres." *BoxOffice* 84, no. 16 (10 February, 1964).

"10 Years of Warner Bros. International Theatres: Promotional Feature." *Screen International* 1150 (20 March, 1998).

"2 Boston-Based Exhibs Offer Upstate N.Y. Sites; Are to Add 24 Screens." *Variety* 314, no. 10 (4 April, 1984).

"6 Reasons Why Hoyts Australia is One of the Top Grossing Cinema Chains in the World!" *Variety* 282, no. 12 (5 May, 1976).

Agnew, J. W. "Cinema + Complex = Cineplex." *BoxOffice* 115, no. 3 (23 April, 1979).

Albarino, Richard. "Just How Many Film Sites Are There?" *Variety* 277, no. 9 (8 January, 1975).

Alderman, Bruce. "Cannon in Holland has 68 Screens & 34% Market Share." *Variety* 332, no. 11 (5 October, 1988).

Alexander, Garth. "WB, Grocer to Build Japan Multiplexes." *Variety* 343, no. 5 (13 May, 1991).

© The Author(s) 2019

S. Hanson, *Screening the World*,

https://doi.org/10.1007/978-3-030-18995-2

Alford, Christopher. "CJ Village at Home in Korean Exhib'n." *Variety* 380, no. 13 (13 November, 2000a).

Alford, Christopher. "Screening for Quality." *Variety* 378, no. 10 (24 April, 2000b).

Alsberge, Harold. "Pathe Emphasizes Customer Satisfaction." *The Film Journal* 97, no. 6 (1 July, 1994).

"AMC Begins Construction on Largest U.S. Multiplex." *The Film Journal* 97, no. 6 (1 July, 1994).

"AMC Bows Nine New Screens in Toledo." *Variety* 320, no. 12 (16 October, 1985).

"AMC Broadens Commitment to 'Aggressive' Expansion." *Screen International* 475 (8 December, 1984).

"AMC Forms Global Arm for Overseas Expansion." *Variety* 317, no. 6 (5 December, 1984).

"AMC in Huge Expansion Plan for 1971 With 70 Auditoriums in 13 Cities." *BoxOffice* 98, no. 26 (12 April, 1971a).

"AMC in Huge Expansion Plan for 1971 With 70 Auditoriums in 13 Cities." 98, no. 26 (12 April, 1971b).

"AMC Opens 7-Theatre Complex in Toledo." *BoxOffice* 10, no. 19 (21 August, 1972).

"AMC Opens Largest Theatre Complex in U.S." *The Film Journal* 98, no. 5 (1 June, 1995).

"AMC Plans Further Expansion in Century City Shopping Centre." *Screen International* 544 (19 April, 1986).

"AMC Quad Announced for Harrisburg, Pa." *BoxOffice* 102, nos. 10–11 (18 December, 1972).

"AMC Target Is 40 Multiplexes by 1990." *Screen International* 525 (30 November, 1985).

"Analysis of UK Multiplex Developments in 1997." *Screen Finance* (5 February, 1998).

Arnold, Andrew. "Convenience Vs. Experience: Millennials Love Streaming But Aren't Ready To Dump Cinema Just Yet." *Forbes* (26 October, 2017). https://www.forbes.com/sites/andrewarnold/2017/10/26/millennials-love-streaming-but-arent-ready-to-dump-cinema-just-yet/#6d6532f46311 (accessed 13 February, 2019).

Attenborough, Richard. "Momentum of British Film Year Can Carry Effects Beyond 1986." *Screen International* 492 (13 April, 1985).

Auerbach, Alexander. "Despite the Christmas-season Chill Circuit Heads Remain Optimistic." *BoxOffice* 118, no. 4 (1 April, 1982).

"Aussie Screens." *Variety* 323, no. 1 (30 April, 1986).

"Awakened Theatre Industry Catching Up With 'Lost' Suburban Patrons." *BoxOffice* 85, no. 11 (6 July, 1964a).

"Awakened Theatre Industry Catching Up With 'Lost' Suburban Patrons." 85, no. 11 (6 July, 1964b).

Baer, Joan. "Enlarges Downtown Activity with a Variety of Operations." *BoxOffice* 81, no. 15 (30 July, 1964).

Bahiana, Ana Maria. "All the World's a Screen." *Screen International* 1047 (1 March, 1996).

Barker, Paul. "The Future Is Here and Now: Lots of Happy Smiling People Tripping to the Shopping Mall. But Does It Work?" *The Guardian*, 8 October, 1996.

Batey, Mark. "Blockbusters and Beyond." *Cinemabusiness* 30 (December, 2005).

Bell, Nick. "Charging Forward." *Screen International* 884 (20 November, 1992).

Beller, Thomas. "The Death of a Movie Theatre." *The New Yorker*, 6 September, 2018.

"Benelux-Films: Power Concentrations Increased Among Belgian Cinema Owners." *Variety* 316, no. 12 (17 October, 1984).

Berggren, G. M. and Leonard, K. R. "Cinema 3 Opening an Historic Event." *BoxOffice* 91, no. 13 (17 July, 1965).

Berliner, Debra. "Cinema Centers Corp. Projects 150 New Screens in Five Years." *The Film Journal* 87, no. 7 (1 July, 1984).

Blaney, Martin. "Germans Fall Before US box-office Advance." *Screen International* 827 (4 October, 1991a).

Blaney, Michael. "Constantin, Warner Plan Multis." *Screen International* 790 (18 January, 1991b).

"Bleep! Zap! Ding! Profits!" *BoxOffice* 117, no. 9 (1 September, 1981).

Blickstein, Jay. "AMC: First Multiplex: An Accident of Design." *Variety* 350, no. 6 (8 March, 1983).

Block, A. B. "What Makes Stanley Borrow? Stan Durwood's AMC Entertainment Is Loaded with Debt and Costly Leases. So Why Does the Stock Sell for 27 Times Earnings?" *Forbes*, 22 September, 1986.

Borger, Lenny. "French Reformers Aim to Split Gaumont, Decentralize Exhibition; Push Local Role for U.S. Majors." *Variety* 305, no. 2 (11 November, 1981).

Borsboom, Edward. *100 Years of Cinema Exhibition in the Netherlands—A Historical Profile*. Media Salles. http://www.mediasalles.it/ybkcent/ybk95_nl.htm (accessed 3 January, 2019).

Brackhurst, Chris. "Multi-screens in Battle with High-Street 'Dinosaurs'." *Sunday Times*, 6 August, 1989.

Bramesco, Charles. "Is Netflix about to Change How It Releases Original Movies?" *The Guardian*, 31 August, 2018.

Brill, Louis M. "Megaplex Rising." *BoxOffice* 130, no. 10 (1 November, 1994).

"Brit. Admissions Up 10% In '83, But Decline Plagues Industry." *Variety* 315, no. 2 (9 May, 1984).

"Britain's Maybox Group to Exit Exhibition: Competish Too Tough." *Variety* 331, no. 6 (1 June, 1988).

"British Admissions Slump; 'Market' by Gut Instinct; A Nation of Home-Owners." *Variety* 301, no. 11 (14 January, 1981).

"British Cinema and Film Statistics." *Screen Digest* (October, 1990).

"British Cinema Fast Disappearing." *Screen Digest* (October, 1984).

"British Cinema Stats." *Variety* 329, no. 13 (20 January, 1988).

Brown, Chris. "British Cinemas Booming, But Can They Keep the Customers Happy?" *Screen International* 159 (7 October, 1978).

Brown, Chris. "West End Linchpin for Cineplex UK Subsidiary." *Screen International* 661 (16 July, 1988).

Brown, Colin. "Carmike Continues Expansion." *Screen International* 1111, no. 6 (June, 1997).

Brown, Colin. "UATG Goes Back on Block." *Screen International* 1147 (27 February, 1998).

Brown, Colin. "Viacom Constructs a 'Global Powerhouse'." *Screen International* 945 (18 February, 1994).

Bryant, Chris. "Vue Pushes into Europe with Cinemaxx Bid." *Financial Times*, 10 July, 2012.

Buckingham, Lisa. "Virgin Sells 'Fantastic' Cinemas for £215m." *The Guardian*, 19 October, 1999.

Buddrus, Petra. "Nordic Giants Jostle for Position." *Screen International* 1222 (20 August, 1999).

Burch, Gary. "The Psychology of Concession Sales; A Look at the Subtlety of Selling." *BoxOffice* 114, no. 24 (19 March, 1979).

Burke-Block, Candace. "M.B.C. Construction Head Cadiff States 'Twinning Is Good, Multiplexing Is Better'." *The Film Journal* 84, no. 4 (9 March, 1981).

Burn-Callander, Rebecca. "Indie Cinema Chain to Challenge UK's 'Homogenous' Giants." *The Daily Telegraph*, 30 June, 2015.

Butler, Robert. "A Night at the Pictures in 1994." *Independent on Sunday*, 16 October, 1994.

Byrne, Bridget. "Exhibition Profile: Making WBIT Tracks: Warner Bros. International Theatres Celebrates its First Decade." *BoxOffice* 134, no. 6 (1 June, 1998).

CAA. *Cinema and Video Industry Audience Research (CAVIAR) 10.* London: CAA, 1993.

CAA. *Cinema and Video Industry Audience Research (CAVIAR) 12.* CAA: London, 1995.

CAA. *Cinema and Video Industry Audience Research (CAVIAR) 20.* London: CAA, 2003.

Campbell, Michael L. "The State of our Art: The Best and Worst of Times: The Head of Regal Cinemas, Newly Crowned King of the Exhibition Universe, Gives Boxoffice's First Annual State of the Industry Address." *BoxOffice* 135, no. 1 (1 January, 1999).

"Candy Bolsters Concession Sales." *BoxOffice* 122, no. 3 (1 March, 1986).

"Cannon to Close Seven City Screens." *Screen International* 661 (16 July, 1988).

Chrisafis, Angelique. "Hoorah for Bollywood as Birmingham Gets Multiplex Screens Dedicated to Indian films." *The Guardian*, 17 May, 2000.

"CIC Plans to Build German Multiplexes." *Variety* 324, no. 4 (20 August, 1986).

"Cinema Concessions: An Essential Luxury." *Screen Digest* (July, 2002).

"Cinema in Europe: Circuit Building and Multiplexes." *Screen Digest* (July, 1991).

"Cinemas and Their Audiences: Just Holding on." *Screen Digest* (September, 1992a).

"Cinemas and Their Audiences: Just holding on." *Screen Digest* (September, 1992b).

"Cineplex Odeon's British Subsid Gallery Hopes to have 110 Screens or more by '91." *Variety* 335, no. 3 (10 May, 1989).

"Cineplex Sells Off 'Non-Core' Business." *Screen International* 750 (31 March, 1990).

Clark, Ted. "Heylen Circuit Renovates Sites and Expands to 60 Screens." *Variety* 311, no. 1 (4 May, 1983).

Clark, Ted. "France to Ape U.S. Divorce, 1950 of Distribs and Theatre Control." *Variety* 306, no. 10 (7 April, 1982).

Clarke, Laurie. "Odeon Cinemas." *Screen International* 891 (22 January, 1993).

Clover, Charles. "Dalian Wanda's AMC to Buy Carmike Cinemas for $1.1bn." *Financial Times*, 4 March, 2016.

CNC (Centre National du Cinéma et de l'Image Animée). *Fréquentation des salles de cinéma: 200,5 millions d'entrées en 2018* (December, 2018). https://www.cnc.fr/professionnels/actualites/frequentation-des-salles-de-cinema%2D%2D2005-millions-dentrees-en-2018_903437 (accessed 13 January, 2019).

"Cobb Praises Christie Equipment in Use at Chain's Cinema City 8." *The Independent Film Journal* 81, no. 13 (1 October, 1978).

"Cobb's 'Octoplex': Eight Theatres, Two Floors of Film, Food and Fun." *BoxOffice* 113, no. 24 (18 September, 1978).

Cohn, Lawrence. "Fewer Plexes but More Multi." *Variety* 341, no. 3 (29 October, 1999).

Combault, Kate. "Cinema Expo Honors UGC Co-founder." *Film Journal International* 103, no. 7 (1 July, 2000a).

Combault, Kate. "Foreign Correspondence: Cinéma Francais." *Film Journal International* 103, no. 6 (1 June, 2000b).

"Construction News: Four AMC Quadruple Complexes Announced." *The Independent Film Journal* 66, no. 8 (16 September, 1970).

"Consuming Movies: Pay TV Eats into Film Spending Cake." *Screen Digest* (January, 1997).

Coopman, Jeremy. "Sell-Through Cheers Labels; Some Execs Wary Of Rental Dip." *Variety* 326, no. 2 (4 February, 1987).

Coopman, Jeremy. "HV Loses No Luster in U.K.; Yanks Clean Up on Rental." *Variety* 335, no. 3 (10 May, 1989).

Cornelius, Andrew. "Film Industry Rallies Against Sale." *The Guardian*, 30 November, 1985.

Cornelius, Andrew. "DTI Clears Cannon Bid Against OFT Advice." *The Guardian*, 14 August, 1986.

Cotterill, Joseph and Ahmed, Murad. "AMC to Buy Odeon from Terra Firma for £921m." *Financial Times*, 12 July, 2016.

Cunningham, John. "Crystal Hall." *The Guardian*, 21 July, 1983.

Daniel, John. "US Challenge to Cinema Chains." *The Guardian*, 14 October, 1988.

"Danish Nordisk Emphasizes Wider Enterprises in Competitive Market." *Variety* 328, no. 13 (12 October, 1987).

Davis Langdon & Everest. "Cost Model." *Building*, 5 May, 2000. http://www.davislangdon.com/upload/StaticFiles/EME%20Publications/CostModels/MultiplexCinemas_CM_5May00.pdf (accessed 12 October, 2011).

Davis, Jonathan and Reeder, Neil. "Chain Reaction." *Screen International* 1019 (4 August, 1995).

Davis, Rebecca. "China Box Office Growth Slows to 9% in 2018. Hits $8.9 Billion." *Variety*, 2 January, 2019. https://variety.com/2019/film/news/china-box-office-2018-annual-1203097545/ (accessed 4 February, 2019).

Dawtrey, Adam. "The Resurrection of British Cinema." *The Hollywood Reporter International Exhibition Special Report* 317 no. 25 (May, 1991).

Dawtrey, Adam. "UGC Pass Piques Plex Players." *Variety* 379, no. 8 (17 July, 2000).

Dinglasan, Francesca. "Operating Megaplexes: Megaplexing Issues." *BoxOffice* 135, no. 9 (1 September, 1999).

Dinglasan, Francesca. "Carmike, Edwards File for Bankruptcy; Regal Next?" *BoxOffice* 136, no. 10 (1 October, 2000a).

Dinglasan, Francesca. "Justice for Reading?" *BoxOffice* 136, no. 1 (1 November, 2000b).

Dinglasan, Francesca. "Reading and No Longer Waiting for First-Run Films." *BoxOffice* 136, no. 12 (1 December, 2000c).

Doward, Jamie. "Multiplexes are Cinemas' Paradiso." *The Observer*, 27 September, 1998.

Downey, Mike. "CIC, UCl Plan 12 German Plexes" *Screen International* 715 (29 July, 1989a).

Downey, Mike. "Multiplex Mania Hits W Germany." *Screen International* 735 (16 December, 1989b).

"Drew Eberson Describes 'Theatre of Tomorrow'." *BoxOffice* 84, no. 2 (4 November, 1963).

Duncan, Celia. "10 Years of UCI: If You Build It, They Will Come." *Screen International* 1026 (22 September, 1995).

"Durwood Plans Three Quads in One Complex." *The Independent Film Journal* 64, no. 9 (30 September, 1969).

"Durwood's the Name!" *Billboard* 52, no. 19 (11 May, 1940).

Edmunds, Marlene. "AMC Plans Swedish 24-plex." *Variety*, 14 October, 1997a. https://variety.com/1997/film/news/amc-plans-swedish-24-plex-1116674243/ (accessed 3 November, 2018).

Edmunds, Marlene. "Hardtop Wars." *Variety* 366, no. 3 (17 February, 1997b).

Edmunds, Marlene. "Kinepolis Keeps the Plexes Coming." *Variety* 371, no. 6 (15 June, 1998a).

Edmunds, Marlene. "Nordic Building Spree Continues." *Variety* 371, no. 3 (15 June, 1998b).

Edmunds, Marlene. "Exhibs Making Inroads." *Variety* 373, no. 12 (8 February, 1999a).

Edmunds, Marlene. "Pathe Plans Dutch Plexes." *Variety* 375, no. 6 (21 June, 1999b).

Elsaesser, Thomas. "The "Return" of 3-D: On Some of the Logics and Genealogies of the Image in the Twenty-First Century." *Critical Inquiry* 39 (Winter, 2013): 217–246.

Ennis-Reynolds, Georgina. "Sustainable Development and Multiplexes." *Journal of Leisure Property*, 2, no. 4 (2002): 317–331.

European Audiovisual Observatory. *Focus 2018: World Film Market Trends.* Marché Du Film: Festival De Cannes, 2018.

"Exhibition on Upswing Globally." *Variety* 320, no. 6 (4 September, 1985).

"Exhibition's First Building Boom in 30 Years Marked by New Designs." 89 (18 July, 1966).

"Exhibs Bid 'Adieu' To Mini-Cinemas." *The Film Journal* 85, no. 18 (23 August, 1982).

Eyles, Allen. "The Virgin Intervention." *Picture House: Magazine of the Cinema Theatre Association*, no. 43 (2018): 8–38.

Faas, Thijs. "Building an Industry." *Screen International* 863 (26 June, 1992).

Fabrikant, Geraldine. "All About/Movie Theaters; The Crunch Comes to Cinemas." *New York Times* (25 November, 1990).

"Far From Resting on Its Laurels, Gaumont, At 90, Launches Prod'n, Distrib and Theater Revamp Plans." *Variety* 326, no. 4 (18 February, 1987).

Faughnder, Ryan and Rottenberg, Josh. "What Netflix's Release of 'Roma' Says About its Movie Business Strategy." *Los Angeles Times*, 3 December, 2018. https://www.latimes.com/business/hollywood/la-fi-ct-netflix-roma-20181203-story.html (accessed 23 December, 2018).

Felperin, Leslie. "Multiplexity." *The PACT Magazine* 56 (September, 1996).

"Filmgoers Flock to Oz' First Multiplex." *Variety* 324, no. 10 (31 December, 1986).

Fithian, John. "Comments by John Fithian President & CEO National Association of Theatre Owners. CinemaCon State of the Industry." Las Vegas, Nevada, 28 March, 2017. https://www.natoonline.org/wp-content/uploads/2017/03/Fithian-CinemaCon-Speech-2017.pdf (accessed 5 July, 2018).

Fleming, Charles. "Snackbar Slowdown Bitter Pill for Exhibs." *Variety* 342, no. 1 (21 January, 1991).

Floyd, Nigel. "View to a Screen Killing." *Stills* (19 May, 1985).

"Focus on Emerging Markets: China." *Screen Digest* (September, 2004).

"Focus on France: Pathé—Keeping Out of Limelight." *Screen International* 445 (12 May, 1984).

"Four-Plex Announced for La Habra, Calif." *BoxOffice* 95, no. 11 (30 June, 1969).

Frater, Patrick. "Pathe Chief Demands Lower Film Rental: Seydoux Unveils Pathe's Multiplex Expansion Plan." *Screen International* 912 (18 June, 1993a).

Frater, Patrick. "UGC Goes to Town on Multiplex Plans." *Screen International* 893 (5 February, 1993b).

Frater, Patrick. "Cinexpo European Focus: France." *Screen International* 963 (24 June, 1994).

Frater, Patrick. "Cinema Expo: Germany." *Screen International* 1013 (23 June, 1995).

Frater, Patrick. "China's Wanda Buys Australia's Hoyts Multiplex Chain." *Variety* (2 June, 2015). https://variety.com/2015/film/asia/chinas-wanda-buys-australias-hoyts-1201510298/ (accessed 23 January, 2019).

Frater, Patrick. "Dadi Pays $575 Million for Orange Sky Golden Harvest's China Cinemas." *Variety* (26 January, 2017a). https://variety.com/2017/biz/asia/dadi-pays-575-million-for-orange-sky-golden-harvests-china-theaters-1201970458/ (accessed 4 February, 2019).

Frater, Patrick. "Wanda Expands Global Theater Reach as AMC Pays $929 Million for Nordic Cinema." *Variety* (23 January, 2017b). https://variety.com/2017/film/finance/wanda-expands-global-theater-reach-as-amc-pays-929-million-for-nordic-cinema-1201966877/ (accessed 12 July, 2018).

Frater, Patrick. "IMAX Signs 30-Theater Deal With China's Jin Yi media." *Variety* (3 April, 2018a). https://variety.com/2018/film/asia/imax-30-theater-deal-with-chinas-jinyi-media-1202742224/ (accessed 7 February, 2019).

Frater, Patrick. "Korea's Lotte Restructures Film Ahead of Overseas Push, Streaming Launch." *Variety* (10 April, 2018b). https://variety.com/2018/biz/asia/korea-lotte-restructures-film-creativeworks-1202748912/ (accessed 29 January, 2019).

Freedman, Peter. "Supermarket Cinema Cashes in." *The Observer*, 8 November, 1987.

Freer, Ian. "Uniplex!" *Empire* 99 (September, 1997).

Frydman, Bruno. "Exporting the Multiplex Model to Europe: The Experience of AMC" at the MEDIA Salles Round Table. *The Impact of Multiplexes on the Cinema Market and on Their Environment*, Amsterdam: Cinema Expo International, 15 June, 1998. http://www.mediasalles.it/expo98fr.htm (accessed 10 September, 2011).

Fuchs, Andreas. "A Multi-screen Pioneer." *Film Journal International* 100, no. 6 (1 July, 1997a).

Fuchs, Andreas. "Tea Houses and Multiplexes." *Film Journal International* 100, no. 1 (1 February, 1997b).

Fuchs, Andreas. "Entertaining Designs: Architect Ira Stiegler Capitalizes on WB Brand." *Film Journal International* 101, no. 12 (1 December, 1998).

Fuchs, Andreas. "Star Entrance." *Film Journal International* 112, no. 2 (February, 2009).

Fuchs, Andreas. "AMC is #1 After Nordic acquisition." *Film Journal* (14 April, 2017). http://www.filmjournal.com/columns/amc-1-after-nordic-acquisition (accessed 2 November, 2018).

George, Sandy. "Greater Union's Mega Hit: 30-Screen Complex Woos Moviegoers in Adelaide Suburb of Marion." *Film Journal International* 101, no. 3 (1 March, 1998).

"German-Speaking Market at a Glance—West Germany." *Variety* 314, no. 6 (7 March, 1984).

Gilbert, W. Stephen. "Foreign Correspondence: The British Picture." *Film Journal International* 103, no. 10 (1 October, 2000).

Glancey, Jonathan. "Slouching Towards Stevenage." *The Independent Magazine*, 7 July, 1996.

Gleason, Larry. "Multiplex Phenomenon Gains Int'l. Momentum." *The Film Journal* 95, no. 2 (1 July, 1992).

"Global Cinema Exhibition Markets." *Screen Digest* (October, 2003).

"Global Cinema Exhibition: Dollar Revenues Fail to Keep up with Soaring Admissions." *Screen Digest* (September, 2002).

"Global Cinema Exhibition: Record Revenues but Dollar Values Mask Local Currency Falls." *Screen Digest* (September, 2004).

Gomer, Hilaire. "Cannon Aims for Multi-Screen Success." *The Guardian*, 24 July, 1985.

Gomery, Douglas. "Hollywood Corporate Business Practice and Contemporary Film History." in *Contemporary Hollywood Cinema*, edited by Steve Neale and Murray Smith. London: Routledge, 1998.

Goodfellow, Melanie. "France Admissions Fell 4.3% in 2018 but it Remains Europe's Top Cinema-Going Territory." *Screen Daily* (3 January, 2019). https://www.screendaily.com/news/france-admissions-fell-43-in-2018-but-it-remains-europes-top-cinema-going-territory/5135539.article (accessed 14 January, 2019).

Goodridge, Mike and Blaney, Martin, "Warner, Village Split in Germany." *Screen International* 1179 (9 October, 1998).

Graser, Marc. "Jolly Times for DVDs." *Variety* 373, no. 5 (14 December, 1998).

Grater, Tom. "Vue Cinemas Acquires Dutch Exhibitor JT Bioscopen." *Screen Daily* (24 August, 2015). https://www.screendaily.com/news/vue-cinemas-acquires-dutch-exhibitor-jt-bioscopen/5091967.article (accessed 3 January, 2019).

"Greater Union Cinema Div. back to High Profile Status." *Variety* 330, no. 13 (20 April, 1988).

Grove, Christopher. Showbiz Runs in the Family. *Variety* 369, no. 7 (22 December, 1997).

Groves, Don. "Cineplex Odeon Sells its Gallery Circuit to Pathé Cannon." *Variety* 338, no. 11 (28 March, 1990a).

Groves, Don. "Yank Cinema Titan UA Selling Blighty Interest to Partner CIC." *Variety* 339, no. 1 (11 April, 1990b).

Groves, Don. "Oz Exhibs Battle Rookie in Suburban Screen Skirmish." *Variety* 365, no. 1 (4 November, 1996).

Guthrie, Jonathon. "Stars Bring Beverly Hills to Birmingham." *Financial Times*, 21 July, 2000.

Halbfinger, David M. "Going Deep for Digital." *New York Times* (26 September, 2005). http://www.nytimes.com/2005/09/26/business/media/26digital.html?pagewanted=all (accessed 20 November, 2017).

"Half of All European Screens Are Multiplexed." *Screen Digest* (November, 2003).

Hall, Katie. "France's Multiplex Gauntlet." *Film Journal International* 100, no. 8 (1 September, 1997).

"Hamilton Approaches Slough with 'Fun'." *Screen International* 628 (28 November, 1987).

Hancock, David. "Digital Cinema—The End of the Beginning, Press Release." *Screen Digest* (25 May, 2005). http://www.screendigest.com/press/releases/FHAN-6CJJDS/pressRelease.pdf (accessed 18 July, 2006).

Hancock, David. "Waving the Magic: Wanda." *Cinema Technology* (December, 2016).

Hancock, David. "Latin America Cinema Markets on the Way up." *Cinema Technology* 30 no. 4 (December, 2017a).

Hancock, David. "Transforming the Movie Experience." *BoxOffice Pro: The Giants of Exhibition 2017* (February, 2017b), 26. https://pro.boxoffice.com/wp-content/uploads/2017/02/Giants2017.pdf (accessed 11 February, 2019).

Harkness, John. "Cineplex Faces More Financial Problems." *Screen International* 390 (16 April, 1983).

Hayes, Graeme. "Regulating Multiplexes: The French State between Corporatism and Globalization." *French Politics, Culture & Society* 23, no. 3 (Winter, 2005): 14–33.

Hazelton, John. "Korff Breaks New Ground." *Screen International* 658 (25 June, 1988).

Hazelton, John. "Drowning in a Sea of Screens." *Screen International* 734 (9 December, 1989).

Heinze, Carolyn. "Today's Lobby: Where Entertainment Begins." *Film Journal International* 103, no. 11 (1 November, 2000).

Henné, Peter. "Dueling Megaplexes, *Film Journal International* 100, no. 5 (1 June, 1997).

Herskovitz, Jon. "Distribs Maintain Hold on Hits Via Archaic Booking Systems." *Variety* 371, no. 2 (18 May, 1998).

Hightower, Susan, "AMC's Grand Multiplex." *Kansas City Star*, 18 May, 1995.

Higson, Andrew. "The Discourses of British Film Year." *Screen* 27, no. 1 (January–February, 1986): 86–109.

Hindes, Andrew. "AMC at 30—Megaplex Dominance Intact." *Variety* 370, no. 5 (16 March, 1998).

Hoad, Phil. "How Multiplex Cinemas Saved the British Film Industry 25 Years Ago." *The Guardian*, 11 November, 2010.

Hodges, Adrian. "650 Jobs Will Go as Rank Closes Cinemas." *Screen International* 298 (27 June, 1981).

Hogarth, John. "Building Admissions Above the 50 Million Mark." *Screen International* 483 (9 February, 1985).

Holberton, Brian. "Today's Lobby." *Film Journal International* 103, no. 11 (1 November, 2000).

Hovatter, David. "Where the Cinema Is Going Wrong." *Screen International* 339 (17 April, 1982).

"Hoyts Chain in Turnaround Via Lift from Sydney Septplex." *Variety* 291, no. 1 (10 May, 1978).

"Hoyts Shoring Up Firm, Cuts Debt Before Next Round of Expansion." *Variety* 335, no. 1 (26 April, 1989).

Ilott, Terry. "Rank Motto: Know When to Say No." *Variety* 343, no. 4 (6 May, 1991).

"In Focus: The Faith of the Decrees." *The Film Journal* 92, no. 6 (1 July, 1989).

"In the Days of the Tent Shows: The Life of the Dubinsky Brothers." *Kansas City Star*, 25 February, 1951.

"Indie Exhib Group Buys French Govt. Circuit (22) For $11,850,000." *Variety* 261, no. 2 (25 November, 1970).

"ITC's Kingham Cites Ills, Possible Aid for Theatres." *Variety* 308, no. 12 (20 October, 1982).

James, Alison. "UGC Pic Pass Raises Gallic Ire." *Variety* 379, no. 5 (19 June, 2000).

Jarvey, Natalie. "Netflix Grows Subscriber Base to 139 Million Worldwide." *The Hollywood Reporter* (17 January, 2019) https://www.hollywoodreporter.com/news/netflix-grows-subscriber-base-139-million-worldwide-1176934 (accessed 12 February, 2019).

Johnson, Greg, Noah, Peter and Martelle, Scott. "O.C. Theater Magnate James Edwards Sr. Dies." *LA Times*, 27 April, 1997. http://articles.latimes.com/1997-04-27/news/mn-53092_1_james-edwards-sr/2 (accessed 1 November, 2018).

"JVC Intros Home Video System." *Backstage* 18, no. 24 (17 June, 1977).

Kaminsky, Ralph. "14-Theatre Beverly Center Cutting Tape This July on 'Cineplex Day'." *The Film Journal* 85, no. 15 (28 June, 1982).

"Kansas City Durwood Circuit to Build Six-Theatre Unit in Omaha Center." *BoxOffice* 92, no. 5 (20 November, 1967).

"Kansas City Twins Are Located Below Ground In Former Storage Area." *BoxOffice* 85, no. 26 (19 October, 1964).

Kawamoto, Mamiko. "T Joy Opens Second E-Cinema Complex in Japan." *Variety* (31 May, 2001) https://www.screendaily.com/t-joy-opens-second-e-cinema-complex-in-japan/405849.article (accessed 30 January, 2019).

Keller, Keith. J.R. "Major Swedish Exhib Chains Undergo Multiplex Conversion." *Variety* 293, no. 9 (3 January, 1979).

Kindred, Jack. "Exhibition Cologne: Ruhring to Go." *Screen International* 823 (6 September, 1991).

Klady, Leonard. "United Artists Theatre Circuit Inc." *Variety* 360, no. 4 (28 August, 1995).

Klady, Leonard. "Megas Lack Plexuality." *Variety* 366, no. 5 (3 March, 1997).

Kollewe, Julia. "Cineworld Buys Picturehouse." *The Guardian*, 6 December, 2012.

Kopper, Phillip. "Movie Theater of the Future: The Home?" *American Film* 3, no. 3 (1 December, 1977): 16–21.

Kramer, Pat. "1995 Giants of Exhibition: Overview: Land of the Giants." *BoxOffice* 131, no. 12 (1 December, 1995a).

Kramer, Pat. "Dutch Treat." *BoxOffice* 131, no. 7 (1 July, 1995b).

Kramer, Pat. "Let the Games Begin!" *BoxOffice* 133, no. 11 (1 November, 1997).

La Franco, Robert. "My Megaplex Is Bigger Than Your Megaplex." *Forbes*, 24 February, 1997. https://www.forbes.com/forbes/1997/0224/5904050a.html#5d8394277a4d (accessed 14 October, 2018).

La Velle, William. "RAC Opens First Over-and-Under Quad and First Six-Theatre Complex." *BoxOffice* 95, no. (21 April, 1969).

Lally, Kevin. "General Cinema Unveils a New Look." *The Film Journal* 88, no. 1 (1 January, 1985).

Lally, Kevin. "New Designs: Exhibitors Strive For Old-Time Luxury." *The Film Journal* 89, no. 6 (1 June, 1986).

Lally, Kevin. "A Smashing British Debut for WB Int'l: Manchester 12-Plex Heralds Ambitious Programme." *The Film Journal* 92, no. 9 (1 October, 1989).

Lally, Kevin and Kelleher, Ed. "Special Report: Megaplex Mania." *Film Journal International* 99, no. 7 (1 August, 1996).

Lambert, Susan. "The Upside from Down Under: A Look at Australia's Booming Exhibition Industry." *BoxOffice* 133, no. 8 (1 August, 1997).

Lane, Harriet. "Fancy a Film Tonight?: Does the Out-of-Town Multiplex Offer More Than Just Pricey Popcorn and Nachos?" *The Observer*, 24 January, 1999.

Lebo, Dina. "New Ideas for Exhibition: Arcade Alternatives." *Film Journal International* 99, no. 5 (1 June, 1996).

Lee, Hyo-won. "South Korea's CJ CGV to Invest $200M in Vietnamese Theaters by 2020." *The Hollywood Reporter*, 9 April, 2017. https://www.hollywoodreporter.com/news/south-koreas-cj-cgv-invest-200m-vietnamese-theaters-by-2020-1034929 (accessed 30 January, 2019).

Linck, David. "Canada's Cineplex: A New Idea Goes South." *BoxOffice* 117, no. 11 (1 November, 1981).

Ludemann, Ralf. "To Protect and Nurture." *Screen International* 863 (26 June, 1992).

Ludemann, Ralf. "Building a Dream Palace: From Site to Sound." *Screen International* 912 (18 June, 1993a).

Ludemann, Ralf. "Building a Dream Palace: Sitting Pretty." *Screen International* 912 (18 June, 1993b).

Ludemann, Ralf. "Exhibition." *Screen International* 891 (22 January, 1993c).

Ludemann, Ralf. "UK Preview Exhibition: Made in America." *Screen International* 1940 (14 January, 1994).

Ludemann, Ralf. "Cine UK Unveils Exhibition Programme." *Screen International* 1011 (9 June, 1995a).

Ludemann, Ralf. "Interview Robert Devereux: Empire Builder." *Screen International* 1019 (3 August, 1995b).

Ludemann, Ralf. "Virgin Territory." *Screen International* 1015 (7 July, 1995c).

Ludemann, Ralf. "Birth of a Notion." *Screen International* 1065 (5 July, 1996a).

Ludemann, Ralf. "National Amusement at 60: On with the Show." *Screen International* 1085 (22 November, 1996b).

Ludemann, Ralf. "German Exhibition: The Space Race." *Screen International* 1095 (14 February, 1997a).

Ludemann, Ralf. "UK Preview 1997: Battle of the Plexes." *Screen International* 1092 (24 January, 1997b).

Magee, Michelle. "Multiplexes Moving in." *Variety* 364, no. 7 (16 September, 1996).

"Major Cinemagoing Countries Collated." *Screen Digest* (October, 1986).

Mann, Jennifer. "Obituary of Stanley Durwood." *Kansas City Star*, 16 July, 1999.

Marich, Robert. "Cannon Special: Focus Cannon Group." *Variety* 332, no. 11 (5 October, 1998).

"Market Profile: European Multiplex Cinemas—The Second Revolution." *Film Journal International* 99, no. 6 (1 July, 1996).

Marsh, Laurie. "The British Exhib's View of the Cinema's Decline." *Variety* 293, no. 9 (3 January, 1979).

Marshall, Colin. "Southdale Center: America's First Shopping Mall." *The Guardian*, 6 May, 2015.

Matthews, Tom. "Theatre Profile: AMC's Century 24." *BoxOffice* 124, no. 2 (1 February, 1988).

Matthews, Tom. "Multiplex Movie Palaces?" *BoxOffice* 124, no. 3 (1 March, 1989a).

Matthews, Tom. "Stanley H. Durwood: The Man Who Invented the Multiplex." *BoxOffice* 125, no. 10 (1 October, 1989b).

Maux Saint Marc, Francoise. "Multiplex Moderation." *Screen International* 1241 (21 January, 2000).

McAsh, Ian. "Take 5: People in Camera." *Films on Screen and Video* 5, no. 6 (June, 1985): 14–15.

McClintock, Pamela. "A 'Fifty Shades' Cocktail? How Booze Is Becoming the New Popcorn at Movie Theaters." *The Hollywood Reporter* (2 February, 2017). https://www.hollywoodreporter.com/news/a-fifty-shades-cocktail-how-booze-is-becoming-new-popcorn-at-movie-theaters-970726    (accessed 27 December, 2018).

McFarling, Tina. "Cinema Exhibition Discussed at BFI Conference Seminar." *Screen International* 516 (28 September, 1985a).

McFarling, Tina. "Survey Points to Customer Dissatisfaction in UK cinemas." *Screen International* 511 (24 August, 1985b).

McFarling, Tina. "Wycombe 6." *Screen International* 610 (25 July, 1987).

McFarling, Tina. "The Plex Factor." *Producer* 12 (Summer, 1990).

McKinney, Clyde R. "Practical Theatre Design: Back to the Future." *BoxOffice* 123, no. 3 (1 March, 1987).

McLendon, Barton R. "Experience of the Recent Past Charts Course for Next Decade." *BoxOffice* 87, no. 12 (12 July, 1965).

"Media Deregulation Is Top Concern Of Senate Committee." *Variety* 301, no. 4 (26 November, 1980).

Meisel, Myron. "Entertainment Centers: Edwards Launches 'the Big One'." *Film Journal International* 99, no. 3 (1 March, 1996).

Meza, Ed. "Link-ups Alleviate German Plex Crunch." *Variety* 379, no. 5 (19 June, 2000).

"Mini-Multi Boom on: See Post-War Theatre Building Peak in 1969." 63, no. 13 (26 May, 1969).

"Mixed Bag For Worldwide Cinema Last Year." *Screen Digest* (February, 2003).

Moran, Mark. "Multiplexes Are a Good Thing." *Cinemabusiness* 20 (December, 2005)

Moran, Mark. "Cinema Showman." *Cinemabusiness* 28 (September, 2006).

"More Harvard Than Hollywood: A Look at GCC's Richard A. Smith." *BoxOffice* 106, no. 24 (24 March, 1975).

Morley, Steven. "Fitting Digital into Theatres: Merging Technology with Existing Operations." *Film Journal International* 103, no. 3 (1 March, 2000).

Morrison, Melissa. "Having a Lark." *BoxOffice* 134, no. 12 (1 December, 1998).

Morrison, Melissa. "Rise of the Netherlands: Vlissingen venture Opens a New Era in Dutch Exhibition." *BoxOffice* 135, no. 4 (1 April, 1999).

"Movie Attendance Hits 11-Year Peak." *BoxOffice* 115, no. 32 (5 November, 1979).

"Movie Theater gets Cut to Size." *Business Week* (14 March, 1970).

"MPAA Reveals US and Canadian Box Office Grosses Down 7% from 1985." *Screen International* 537 (1 March, 1986).

"Multiplex Cinemas: Growth Is Far From Over." *Screen Digest* (February, 1995).

Murdoch, Blake. "A Greater Union: Leading Aussie Circuit Plans to Double its Screen Count." *Film Journal International* 100, no. 1 (1 February, 1997).

Murdoch, Blake. "Hoyts, CIC Buddy up to Spread Multiplexes Through the Suburbs." *Variety* 324, no. 3 (13 August, 1986).

Murphy, A. D. "Big Film Trend Stories of '73." *Variety* 273, no. 9 (9 January, 1974).

Murray, Karen. "Cineplex Odeon 15th Anniversary—Exhibber Back from the Brink." *Variety* 355, no. 2 (25 July, 1994).

"N.Y. Now Cinema Centers' Top Market; Adding Albany Ten-Plex." *Variety* 310, no. 20 (20 April, 1983).

Natale, Richard. "Designed to Entertain." *Variety* 373, no. 7 (4 January, 1999).

"NATO Okays Distrib Exhibs." *The Film Journal* 89, no. 9 (1 September, 1986).

Neckebroeck, Kjell. "France and Benelux: The Exhibition Basics." *The Film Journal* 99, no. 1 (1 January, 1996).

"News—British Film Year." *AIP & Co.* 59 (October, 1984).

Noglows, Paul. "Here Come the Megaplexes: Exhibs Usher in 24-screen 'Destinations'." *Variety* 356, no. 4 (22 August, 1994).

Ochs, Millard. "Plexing Muscle." *Screen International* 1150 (20 March, 1998).

Olins, Wally. "The Best Place to See a Film?" *Sight and Sound* 54, no. 4 (Autumn, 1985): 241–244.

"The Operators." *Cinemabusiness* (December, 2005).

Park, James. "U.K. Circuit Expansion, Renovation Thought to Be Cure for Ailing Biz." *Variety* 316, no. 10 (3 October, 1984).

"Parkway One, Two Adds Paris Touch to Shop Center's Roof Promenade." *BoxOffice* 83, no. 4 (20 May, 1963).

Phillips, Braden. "Regal at 10: Stadium Seating Tops List of Theater Design Priorities." *Variety* 374, no. 3 (8 March, 1999).

Pitman, Jack. "About One-Third of ABC Cinemas No Longer Viable, Asserts Lennox." *Variety* 219, no. 2 (8 May, 1985).

"The Point—Signalling a New Era in Cinema Exhibition?" *AIP & Co.*, 78 (October, 1986).

Porter, Beth. "Virgin Exhibitors." *The Film Journal* 98, no. 9 (1 December, 1995).

Porter, Beth. "Cine-UK's Wiener Challenges Big Exhibition Guns." *The Film Journal* 99, no. 1 (1 January, 1996).

Porter, Beth. "New England Comes to England: National Amusements Continues Its United Kingdom Expansion." *Film Journal International* 100, no. 1 (1 February, 1997a).

Porter, Beth. "UCI Looks East." *Film Journal International* 100, no. 1 (1 February, 1997b).

Pristin, Terry. "Loews Cineplex May Seek Bankruptcy Protection." *New York Times* (24 January, 2004).

Proctor, Bruce. "Building for the '90s Includes Better Concession Facilities." *The Film Journal* 96, no. 2 (1 March, 1993).

Pryor, Thomas. M. "The New Exhibition Mania." *Variety* 32, no. 2 (6 August, 1986).

"Puttnam Opposed to Cannon Buy of ABC Chain as Duopoly Reprise." *Variety* 323, no. 4 (21 May, 1986).

"Puttnam's Applause." *The Citizen*, 28 November, 1985.

Quinn, John W. "American Multi at 457; Formula: Follow Malls." *Variety* 285, no. 8 (29 December, 1976).

"RAC Breaks Ground for Houston Six." *BoxOffice* 94, no. 26 (14 April, 1969).

"Rank Aims for 200 Odeon Screens in 75 Locations with Expansion Plans." *Screen International* 601 (23 May, 1987).

"Reading Enters Oz." *The Film Journal* 99, no. 2 (1 February, 1996).

"A Redstone Milestone." *Film Journal International* 99, no. 1 (1 November, 1996).

Redstone, Sumner. "10 Years of Showcase Cinemas: Building Blocks—Promotional Feature." *Screen International* 1162 (12 June, 1998).

"Reformation of UK Film Exhibition." *Screen Digest* (February, 1984).

"Regal At 10: Regal Cinemas Corporate History." *Variety* 374, no. 3 (8 March, 1999).

Reich, Volker. "A Multiplex in Every large City." *The Film Journal* 98, no. 6 (1 July, 1995).

Reich, Volker. "Booming Investments Vs. Laisser-Faire in Germany." *Film Journal International* 100, no. 6 (1 July, 1997).

Robins, Jim. "Exhibs See a Trend to Larger Houses: Big Screen to Fight Homevideo." *Variety* 308, no. 1 (4 August, 1982).

Robbins, Jim. "Redstone Bares Strategy in Building Circuit Empire." *Variety* 323, no. 11 (9 July, 1986).

Rosenberg, Scott. "AMC Comes to Asia: U.S. Circuit Finds Success in Far East." *Film Journal International* 102, no. 3 (1 March, 1999).

Rosenberg, Scott. "Reason to Smile." *Film Journal International* 100, no. 4 (1 May, 1997).

Rossing Jenson, Jorn. "Environmentally Sound." *Screen International* 734 (9 December, 1989).

Rossing Jenson, Jorn. "Swedish Distrib Svensk Snubs New AMC Multiplex." *Variety* 382, no. 8 (11 April, 2001).

Rouse III, Charles, F. "Will Faltering Economy Curtail Theater Growth?" *BoxOffice* 116, no. 18 (5 May, 1980).

Rubin, Rebecca. "Record Breaker! Domestic Box Office Hits New Year High in 2018." *Variety* (24 December, 2018). https://variety.com/2018/film/news/domestic-box-office-record-2018-1203095448/ (accessed 28 December, 2018).

Salisch, Wynn. J. "The Impact of Programming on Concession Sales." *Film Journal International* 102, no. 11 (1 November, 1999).

Sandy, George. "Readings in Australia." *Film Journal International* 101, no. 4 (1 April, 1998).

Schilling, Mark. "UCI Puts Smile into Chinese Exhibition." *Screen International* 952 (8 April, 1994).

Schutter, Christian de. "Think Big." *Screen International* 734 (9 December, 1989).

Scott, Mary. "Intercity Express." *Screen International* 1131 (24 October, 1997).

Scott, Mary. "Virgin Pulls UK-Irish Exhib Plug." *Screen International* 1231 (22 October, 1999a).

Scott, Mary. "Executive Suite: Dave Harris." *Screen International* 1273 (25 August, 2000a).

Scott, Mary. "StarCity Highlights Megaplex Concerns." *Screen International* 1269 (28 July, 2000b).

Scott, Mary. "UGC Card Provokes UK Distribution Fury." *Screen International* 1270 (4 August, 2000c).

Scott, Russell. "Exporting Expertise." *Film Journal International* 102 no. 7 (1 July, 1999b)

Segers, Frank. "Say VCR Effect on TIX Sales Peaking." *Variety* 321, no. 12 (15 January, 1986).

Selig, Robert E. "Exhibs Think Big as Filmgoers Have Small Home Screens." *Variety* 321, no. 11 (8 January, 1986).

Sexton, David. "The Empire Strikes Back." *Sunday Telegraph Magazine*, 12 March, 1989.

Simms, Calvin. "Synergy': The Unspoken Word." *New York Times* (5 October, 1993).

Skodack, Debra. "Going Grand in Dallas, *Kansas City Star*, 9 June, 1994.

Smith, Justin. "Cinema for Sale: The Impact of the Multiplex on Cinema Going in Britain, 1985–2000." *Journal of British Cinema and Television* 2, no. 2 (2005): 242–255.

Smith, Richard A." Abundant Potential for Progress in Shopping Center Theatres." *BoxOffice* 87, no. 12 (12 July, 1965).

Smith, Roger. "The Street Mulls Megaplex Mania." *Variety* 377, no. 5 (13 December, 1999).

Sohee, Kim. "Foreign Correspondence: Korea Scene." *Film Journal International* 101, no. 4 (1 April, 1998).

"Sony Lincoln Square Theatre Provides View of the Future." *The Film Journal* 98, no. 22 (1 March, 1995).

"Special Report: Strategic Trends in Theatrical Exhibition: The Independent Exhibitor's Survival Guide." *BoxOffice* 124, no. 7 (1 July, 1989).

Spangler, Todd. "Amazon's Prime Video Channels Biz to Generate $1.7 Billion in 2018." *Variety* (7 December, 2018) https://variety.com/2018/digital/news/amazon-prime-video-channels-tv-revenue-estimates-1203083998/ (accessed 13 February, 2019).

Steen, Al. "Twin Construction May Start a New Trend." *BoxOffice* 81, no. 20 (3 September, 1962).

"Step Into the Future With Us." *Variety* 371, no. 6 (15 June, 1998).

Stern, Andy. "Kinepolis Still Rules the Belgian Scene." *Variety* 375, no. 6 (21 June, 1999).

Stern, Andy. "UGC, Kinepolis Duke it Out for Belgian Dominance." *Variety* 379, no. 5 (19 June, 2000).

"Stockholm's Filmstaden Adds Four New Cinemas, Aims for a Million Admissions." *Screen International* 471 (10 November, 1984).

Strub, Kimberley. "The European Film and Exhibition Climate." *BoxOffice* 128, no. 6 (1 June, 1992).

"Suit to Durwood: Referee in 20 million-Dollar Action Rules for the Theater Executive." *The Kansas City Star*, 21 December, 1953.

Summers, Jimmy. "Cineplex Odeon Launches a Dazzling New Flagship." *BoxOffice* 123, no. 9 (1 September, 1987).

Summers, Sue. "'Save UK Cinema' Plea to State." *Screen International* 98 (30 June, 1977).

Sutherland, Alex. "AMC Unveils Cinema Investments Plan as UK Admissions Show Big Gains." *Screen International* 492 (13 April, 1985).

Sweeney, Marje. "New Academy Theatre Joins Galaxy of Durwood Downtown Showcases." *BoxOffice* 88, no. 9 (17 December, 1962).

Tagliabue, John. "Now Playing Europe: Invasion of the Multiplex; With Subplots on Pride and Environment." *The New York Times* (27 January, 2000).

Taylor, Paul. "US Consumer Electronics Sales to Top $10bn." *Financial Times*, 8 January, 2004.

"Ten Giants: Carmike." *BoxOffice* 124, no. 12 (1 December, 1988).

"Ten Giants: Hoyts." *BoxOffice* 124, no. 12 (1 December, 1988).

"Ten Giants: United Artists." *BoxOffice* 124, no. 12 (1 December, 1988).

"The Film Works for UCI Cinemas." *Designweek*, 1 September, 2000. https://www.designweek.co.uk/issues/31-august-2000/the-film-works-for-uci-cinemas/ (accessed 20 November, 2018).

"Theatres in Shopping Centers Can Change the Entertainment Industry." *Back Stage* 2, no. 30 (25 August, 1961).

"Three Pact to Build Multiplexes in the Netherlands." *Variety* 351, no. 2 (10 May, 1993).

Toumarkine, Doris. "In the European Avant-Garde: CGR Cinemas Pioneers Digital Projection in France." *Film Journal International* 111, no. 7 (18 June, 2008).

Townsend, Matt, Surane, Jenny, Orr, Emma, and Cannon, Christopher. "America's 'Retail Apocalypse': Is Really Just Beginning." *Bloomberg* (8 November, 2017), https://www.bloomberg.com/graphics/2017-retail-debt/ (accessed 23 December, 2018).

Trainer, David. "Reality Is Closing in on Netflix." *Forbes* (5 February, 2019) https://www.forbes.com/sites/greatspeculations/2019/02/05/reality-is-closing-in-on-netflix/#1c25cfc5ed0b (accessed 13 February, 2019).

Trout, Wesley. "Lively Interiors Enhance New Theatre." *BoxOffice* 85, no. 16 (10 August, 1964).

Turnbull, Dale. "Australia Rebuilding; Theatres in Plazas Share Lobby Like in U.S." *Variety* 253, no. 8 (8 January, 1969).

Turnbull, Dale. "Confidence: It Builds Theatres." *Variety* 257, no. 8 (7 January, 1970).

Tusher, Will. "General Cinema Plans to Add 300." *Variety* 317, no. 6 (5 December, 1984).

Tusher, Will. "Justice Dep't. Has Eye on Distributors: Collusion's What it's Watching For." *Variety* 324, no. 2 (6 August, 1986).

"U.S. and Canada theatre Circuits Ranked By Size." *Variety* 338, no. 4 (31 January, 1990).

"UCI Opening 8-plex, London's first multi." *Variety* 337, no. 7 (22 November, 1989).

"Uneven Pace of European Cinema." *Screen Digest* (September, 2001).

"United Kingdom Cinema Market." *Screen Digest* (July, 1991).

Urban, Andrew L. "Cracking the Nut." *Screen International* 734 (9 December, 1989), 24.

"US District Court Southern District of New York. Text of the Opinion of the Statutory Court in the Antitrust Suit In the Litigation Against Remaining Defendants, 20th Century Fox, Loew's Inc. and Warner Bros." *BoxOffice* 55, no. 13 (30 July, 1949).

Valenti, Jack. "Theatre Owners and Changing Audience." *Screen International* 285 (28 March, 1981).

Van De Kerkof-Flanagan, Mary. "Dutch Theater Boom Continues." *Variety* 356, no. 13 (24 October, 1994).

Vanier, Fiona. "Multiplex Numbers Fall for First Time in 2002." *Screen Finance* 16, no. 6 (26 March, 2002).

"Varied Array of Firstruns and Art Pictures Fills Century City 14." *Variety* 328, no. 12 (14 October, 1986).

"Village Roadshow Finds Middle Titles Also Getting Results." *Variety* 327, no. 1 (29 April, 1987).

"Virgin Territory." *Film Journal International* 101, no. 7 (1 July, 1998).

von Sychowski, Patrick. "Every Multiplex Is New Again." *Cinema Technology* 31, no. 1 (March, 2018).

Waldman, Alan. "The Road to Asia." *The Hollywood Reporter Salute to Showester of the Year Stan Durwood* (4 March, 1996).

"Warner Village's Rising Star." *Screen International* 1269 (28 July, 2000).

Watkins, Roger. "Cannon Grabs Screen Entertainment." *Variety* 323, no. 2 (7 May, 1986).

Watts, Anita. "Attendance and Concessions: We Need Movies! Snack Corner Part II." *Film Journal International* 103, no. 11 (1 November, 2000).

"WB: Joint Venture, Per Capita." *BoxOffice* 127, no. 7 (1 July, 1991).

Weisman, Milton C. "The Paramount Decrees." *The Independent Film Journal* (30 June, 1956).

"What to Do About U.K. Cinemas, Now the Worst off in Europe." *Variety* 317, no. 12 (16 January, 1985).

Widem, Allen M. "New Shopping Center, Theatre Construction Trends Toward Expansion on Existing Sites." *BoxOffice* 109, no. 17 (2 August, 1976).

Wilkinson, Carl. "Welcome to Cinema Paradiso." *The Observer*, 17 August, 2003.

"Will Automation Revolutionize the Industry?" 95, no. 26 (13 October, 1969).

Williams, Michael. "Indie Exhibs Fear Pathe Swap Fallout." *Variety* 246, no. 1 (20 January, 1992).

Williams, Michael. "Plexes Pack 'em in, Franco Study Finds." *Variety* 369, no. 2 (17 November, 1997).

Williams, Michael. "Gauls to AMC: Yankee Go Home!" *Variety* 370, no. 10 (20 April, 1998).

Williams, Michael and Hils, Miriam. "Local Euro Pix: Big Haul at Mall." *Variety* 366, no. 6 (10 March, 1997).

Williamson, Kim. "Yankee Go Home?" *BoxOffice* 130, no. 7 (1 July, 1994).

"With Cineplex Plus Odeon, Garth Drabinsky Adds Up To Major Theater Force." *Variety* 317, no. 4 (21 November, 1984).

"World Cinema: Upward Reversal of Fortune." *Screen Digest* (September, 1994).

"World VCR Markets: Back Into Growth." *Screen Digest* (June, 1993).

"World's First Six-Theatre Complex Opens in Omaha Under ARC Banner." *BoxOffice* 94, no. 15 (27 January, 1969).

"Worldwide Cinema: Key Markets Strong as Global Admissions Dip." *Screen Digest* (October, 2001).

"Worldwide Cinema: Poor Product Fails the Multiplex Boom." *Screen Digest* (September, 2000).

"Worldwide Cinema: Weakness Hidden by Record Growth." *Screen Digest* (September, 1998).

"Worldwide DVD Markets in Position." *Screen Digest* (November, 2004).

Wroe, Martin. "Pre-Packed Fun in the Pleasure Dome." *The Observer*, 8 October, 1995.

Younge, Gary. "The Big Picture." *The Guardian*, 26 July, 2000.

Zeitchik, Steven. "Is Netflix Killing the Movie Theater? Not So Fast." *The Washington Post*, 24 December, 2018. https://www.washingtonpost.com/business/is-netflix-killing-the-movie-theater-not-so-fast/2018/12/24/7a16dbf8-037a-11e9-8186-4ec26a485713_story.html?noredirect=on&utm_term=.273d927a5d93 (accessed 28 December, 2018).

# Index[1]

---

[1] Note: Page numbers followed by 'n' refer to notes.

© The Author(s) 2019
S. Hanson, *Screening the World*,
https://doi.org/10.1007/978-3-030-18995-2

The manufacturer's authorised representative in the EU is Springer
Nature Customer Service Centre GmbH, Europaplatz 3, 69115 Heidelberg,
Germany. If you have any concerns regarding our products, please
contact ProductSafety@springernature.com

Printed and bound by CPI Group (UK) Ltd, Croydon, CR0 4YY
23/04/2026
02095601-0006